TEST BANK

APPLIED COLLEGE ALGEBRA

A Graphing Approach

by Gareth Williams

Henry Smith
Southeastern Louisiana University

Saunders College Publishing
A Division of Harcourt College Publishers

Fort Worth Philadelphia San Diego New York Orlando Austin
San Antonio Toronto Montreal London Sydney Tokyo

Printed in the United States of America

ISBN 0-03-026041-8

012 023 7654321

Preface

This Test Bank is a supplement to accompany *Applied College Algebra: A Graphing Approach* by Gareth Williams. It contains over 1600 questions arranged by chapter in prepared test format. For each chapter there are 6 tests, labeled Forms A-F, of 25 questions each. Forms A-C are in multiple-choice format; forms D-F have open-ended questions. Four final exams, 2 each in multiple-choice and open-ended format, of 40 questions each, are also included.

Answers to all questions are given at the end of this manual. The answer section also includes reference information about the section, page number, and difficulty level for each problem. There are three levels of difficulty: 1 is the hardest, 3 is the easiest. For example, the answer for Chapter 4, Form D, question 3, gives the reference information "[4.1/327/3]" to indicate that the problem comes from Section 4.1, page 327, and is of difficulty level 3 (easiest).

This Test Bank is also available on CD-ROM for Windows® and Macintosh® formats. The ESATEST 2000™ Computerized Test Bank CD-ROM allows you to preview, select, edit, and add items to tailor tests to your course or select items at random; add or edit graphics, and print up to 99 different versions of the same test and answer sheet. The On-Line Testing feature allows you to create tests and administer them over a network. These tests can then be scored and the grades recorded using the ESAEST gradebook software. If you are interested in receiving a copy of the ESATEST 2000™ Computerized Test Bank CD-ROM or would like more details about its features, please contact your local Saunders College Publishing representative.

Special thanks to Daniel Biles of Western Kentucky University for his careful checking of the accuracy of all the items in this test bank. His help is greatly appreciated. If you have any comments, suggestions, or corrections for this test bank, please address your correspondence to: Mathematics Editor, Saunders College Publishing, Public Ledger Building, Suite 1250, 150 South Independent Mall West, Philadelphia, PA 19106.

Test form RA Date __________ Name ______________________________
Course ______ Sec ______ Student Number __________________

Chapter R

R-1. Let $X = \left\{-5, -\frac{7}{5}, 0, \sqrt{7}\right\}$. List all the elements of X that are rational numbers.

A. $-5, -\frac{7}{5}$

B. $\sqrt{7}$

C. $-5, -\frac{7}{5}, 0$

D. -5

E. $-5, -\frac{7}{5}, \sqrt{7}$

R-2. Indicate which algebraic property of real numbers is used to justify the statement $2x + 3y = 3y + 2x$.

A. Commutative property of multiplication

B. Commutative property of addition

C. Associative property of multiplication

D. Associative property of addition

E. Distributive property

R-3. Compute the expression $|5.02 - 8.14| \div 3.12$ to four decimal places.

A. -1

B. 1

C. -3.1

D. 3.1

E. 2.4110

R-4. Evaluate $\sqrt[4]{(-5)^4}$

A. -5

B. 5

C. 2.2361

D. -2.2361

E. 25

R-5. Use the rules of exponents to write the algebraic expression in a form that involves each variable at most once to a positive exponent.

$$\frac{4^3x^2y^7}{2^3x^5y^4}$$

A. $\frac{8y^3}{x^3}$ B. $8x^{2/5}y^{7/4}$ C. $\frac{y^3}{8x^3}$

D. $8x^3y^3$ E. $\frac{8x^3}{y^3}$

R-6. Simplify $\sqrt[4]{3x^4y^6}$ by removing as many factors as possible from under the radical.

A. $xy\sqrt[4]{3y^2}$ B. $|xy|\sqrt[4]{3y^2}$ C. $xy\sqrt{3y}$

D. $3xy\sqrt{y}$ E. $-xy\sqrt{3y}$

R-7. Use a calculator to compute $4.01^{-3.7}$. Express the result to four decimal places.

A. 170.4646 B. 170.4651 C. 0.0059

D. 0.0072 E. 0.0081

R-8. Give the degree of the polynomial. $12x^4+5x^2-9$

A. 3 B. 12 C. 4 D. 16 E. 17

R-9. Perform the operations of addition and subtraction of polynomials.
$(5x^2+3x+7)+(4x^2+7x+4)-(3x^2+5x+7)$

A. $12x^2+15x+18$ B. $6x^2-15x-18$

C. $6x^2+5x+4$ D. $6x^2+15x+18$

E. $6x^2-5x-4$

R-10. Multiply the polynomials. $(2x^3+5)(3x+4)$

A. $6x^4+8x^3+15x+20$

B. $6x^2+23x+20$

C. $6x^4+23x^2+20$

D. $6x^4+20$

E. $6x^2+20$

R-11. Divide x^2+4x+7 by $x+2$.

A. $x+1+\dfrac{7}{x+2}$

B. $x+2+\dfrac{3}{x+2}$

C. $x+2+\dfrac{7}{x+2}$

D. $x+1+\dfrac{3}{x+2}$

E. $x+2-\dfrac{3}{x+2}$

R-12. Factor $x^2+2x-24$

A. $(x-4)(x-6)$

B. $(x+4)(x-6)$

C. $(x-4)(x+6)$

D. $(x-24)(x-1)$

E. $(x-24)(x+1)$

R-13. The expression is the difference of two squares. Factor $25x^2-(3y+1)^2$.

A. $(25-3y+1)(25+3y+1)$

B. $(5x-3y-1)(5x+3y+1)$

C. $(5x+3y+1)^2$

D. $(25x-3y+1)(25x+3y+1)$

E. $(25x-3y+1)(25x+3y+1)$

R-14. Factor the perfect square $4x^2 - 20x + 25$.

A. $(2x-5)^2$ B. $(2x+5)^2$ C. $(2x+1)^2$

D. $(2x-1)^2$ E. $(2x-25)^2$

R-15. Simplify the rational expression. $\dfrac{(x^2+x-6)}{(x^2+3x-10)}$

A. $\dfrac{x+3}{x+5}$ B. $\dfrac{x-3}{x-5}$ C. $\dfrac{x-6}{x-10}$

D. $\dfrac{x+6}{x+10}$ E. $\dfrac{x^2+x-6}{x^2+3x-10}$

R-16. Perform the multiplication. $\dfrac{(5x+4)^2}{25} \cdot \dfrac{10}{5x+4}$

A. $5x+4$ `B. $2(x+4)$ C. $2(5x+4)$

D. $25(5x+4)$ E. $\dfrac{2(5x+4)}{5}$

R-17. Perform the indicated operations. $\dfrac{x^2-4x-21}{x^2+2x-3} \cdot \dfrac{x^2+6x-7}{x^2-49}$

A. 1 B. $\dfrac{x-3}{x+3}$ C. $\dfrac{x+7}{x+3}$

D. $\dfrac{x-7}{x+7}$ E. $\dfrac{x+7}{x-7}$

R-18. Perform the indicated operation. $\dfrac{6}{x-1}+\dfrac{7}{3x+1}$

A. $\dfrac{13}{4x}$

B. $\dfrac{1}{(x-1)(3x+1)}$

C. $\dfrac{25x-1}{(x-1)(3x+1)}$

D. $\dfrac{25x+1}{(x-1)(3x+1)}$

E. $\dfrac{-1}{(x-1)(3x+1)}$

R-19. Solve the equation for x. $4x+5=7$

A. 3

B. $\dfrac{1}{2}$

C. 2

D. −3

E. −2

R-20. Solve the equation for x. $5(x+3)=10(2x+1)$

A. $\dfrac{1}{3}$

B. 3

C. −3

D. $-\dfrac{1}{3}$

E. 0

R-21. Solve the equation for x, if a solution exists. $\dfrac{x}{x-2}-\dfrac{2}{x-2}=4$

A. 2

B. 0

C. 1

D. −2

E. No solution

R-22. If $f(x)=2x^2+5$, find $f(3)$.

A. 0

B. 23

C. 15

D. 7

E. −3

R-23. Find the domain of $f(x) = \frac{1}{\sqrt{x-4}}$.

A. $(4, \infty)$ B. $[4, \infty)$ C. $(-\infty, 4)$

D. $(-\infty, 4]$ E. $(-\infty, -4)$

R-24. Compute $f(-8.7)$ for $f(x) = 2.01x^2 + 4.21x + 2.17$ to four decimal places.

A. 117.9111 B. 117.6799 C. 118.0112

D. 117.7201 E. 117.8142

R-25. Determine the difference quotient $\frac{f(x+h)-f(x)}{h}$, $h \neq 0$ for $f(x) = 2x^2 + 3$. Give answer in simplified form.

A. $4hx + 2h$ B. $4x + 2h^2$ C. $4x$

D. $4x + h$ E. $4x + 2h$

Test form RB Date _________ Name _______________________

Course ______ Sec ______ Student Number _______________

Chapter R

R-1. Let $X = \left\{-4, -\frac{4}{3}, 0, \sqrt{8}\right\}$. List all the elements of X that are rational numbers.

A. $-4, -\frac{4}{3}, 0$ B. $\sqrt{8}$

C. $-4, -\frac{4}{3}$ D. -4

E. $-4, -\frac{4}{3}, \sqrt{8}$

R-2. Indicate which algebraic property of real numbers is used to justify the statement $4x + 9y = 9y + 4x$.

A. Commutative property of multiplication

B. Commutative property of addition

C. Associative property of multiplication

D. Associative property of addition

E. Distributive property

R-3. Compute the expression $|6.04 - 9.13| \div 2.15$ to four decimal places.

A. -3.09 B. 3.09 C. -1.4372

D. 1.4372 E. 1.7935

R-4. Evaluate $\sqrt[6]{(-7)^6}$

A. -7 B. 1.9129 C. 7

D. 2.6458 E. -1.9129

R-5. Use the rules of exponents to write the algebraic expression in a form that involves each variable at most once to a positive exponent.

$$\frac{4^3x^3y^8}{2^3x^6y^5}$$

A. $\dfrac{8y^3}{x^3}$ B. $8x^{2/5}y^{7/4}$ C. $\dfrac{y^3}{8x^3}$

D. $8x^3y^3$ E. $\dfrac{8x^3}{y^3}$

R-6. Simplify $\sqrt[6]{5x^6y^8}$ by removing as many factors as possible from under the radical.

A. $xy\sqrt[6]{5y^2}$ B. $|xy|\sqrt[6]{5y^2}$ C. $xy\sqrt{5y}$

D. $-xy\sqrt{5y}$ E. $5xy\sqrt{y}$

R-7. Use a calculator to compute 5.2^{-43}. Express the result to four decimal places.

A. 1198.9867 B. 1198.9873 C. 0.0008

D. 0.0012 E. 0.0002

R-8. Give the degree of the polynomial. $15x^4 + 6x^2 - 9$

A. 15 B. 4 C. 3 D. 19 E. 20

R-9. Perform the operations of addition and subtraction of polynomials.
$(9x^2 + 7x + 11) + (4x^2 + 7x + 4) - (3x^2 + 5x + 7)$

A. $10x^2 - 9x - 8$ B. $16x^2 + 19x + 22$

C. $10x^2 + 9x + 8$ D. $16x^2 + 19x + 22$

E. $10x^2 - 19x - 22$

R-10. Multiply the polynomials. $(3x^3+7)(2x+4)$

A. $6x^2+11x+28$

B. $6x^4+28$

C. $6x^4+11x^2+28$

D. ex^2+28

E. $6x^4+12x^3+14x+28$

R-11. Divide x^2+8x+2 by $x+2$.

A. $x+6+\dfrac{-10}{x+2}$

B. $x+6+\dfrac{14}{x+2}$

C. $x+6+\dfrac{-14}{x+2}$

D. $x+10+\dfrac{22}{x+2}$

E. $x+10+\dfrac{18}{x+2}$

R-12. Factor $x^2+2x-15$.

A. $(x-3)(x+5)$

B. $(x+3)(x-5)$

C. $(x-3)(x-5)$

D. $(x-15)(x-1)$

E. $(x-15)(x+1)$

R-13. The expression is the difference of two squares. Factor $36x^2-(3y+1)^2$.

A. $(6x-3y-1)(6x+3y+1)$

B. $(6x-3y+1)(6x+3y+1)$

C. $(6x+3y+1)^2$

D. $(36x-3y+1)(36x+3y+1)$

E. $(36-3y+1)(36+3y+1)$

R-14. Factor the perfect square $9x^2 - 24x + 16$.

A. $(3x+4)^2$ B. $(3x-4)^2$ C. $(3x-16)^2$

D. $(3x-1)^2$ E. $(3x+1)^2$

R-15. Simplify the rational expression. $\dfrac{(x^2+x-12)}{(x^2+2x-15)}$

A. $\dfrac{x-4}{x-5}$ B. $\dfrac{x+4}{x+5}$ C. $\dfrac{x-12}{x-15}$

D. $\dfrac{x+12}{x+15}$ E. $\dfrac{x^2+x-12}{x^2+2x-15}$

R-16. Perform the multiplication. $\dfrac{(6x+5)^2}{25} \cdot \dfrac{10}{6x+5}$

A. $6x+5$ B. $2(x+5)$ C. $2(6x+5)$

D. $\dfrac{2(6x+5)}{5}$ E. $25(x+5)$

R-17. Perform the indicated operations. $\dfrac{x^2-7x+12}{x^2-4x+3} \cdot \dfrac{x^2+3x-4}{x^2-16}$

A. 1 B. $\dfrac{x-3}{x+3}$ C. $\dfrac{x-4}{x+4}$

D. $\dfrac{x-3}{x-4}$ E. $\dfrac{x+4}{x-4}$

R-18. Perform the indicated operation. $\dfrac{5}{x-1}+\dfrac{8}{3x+1}$

A. $\dfrac{13}{4x}$

B. $\dfrac{1}{(x-1)(3x+1)}$

C. $\dfrac{23x-3}{(x-1)(3x+1)}$

D. $\dfrac{23x+3}{(x-1)(3x+1)}$

E. $\dfrac{-1}{(x-1)(3x+1)}$

R-19. Solve the equation for x. $3x+9=2$

A. $-\dfrac{7}{3}$

B. $\dfrac{7}{3}$

C. -7

D. 7

E. $\dfrac{11}{3}$

R-20. Solve the equation for x. $5(x+3)=10(2x+1)$

A. $\dfrac{1}{3}$ B. 3 C. -3 D. $-\dfrac{1}{3}$ E. 0

R-21. Solve the equation for x, if a solution exists. $\dfrac{x}{x-3}-\dfrac{3}{x-3}=4$

A. 3

B. 0

C. 1

D. -3

E. No solution

R-22. If $f(x) = 4x^2 + 3$, find $f(2)$.

A. 0 B. 13 C. 19

D. 7 E. 1

R-23. Find the domain of $f(x) = \dfrac{1}{\sqrt{x-10}}$.

A. $(10, \infty)$ B. $[10, \infty)$ C. $(-\infty, 10)$

D. $(-\infty, 10]$ E. $(-\infty, -10)$

R-24. Compute $f(-4.7)$ for $f(x) = 3.02x^2 + 5.81x + 3.15$ to four decimal places.

A. 42.5419 B. 42.5493 C. 42.5485

D. 42.5471 E. 42.5548

R-25. Determine the difference quotient $\dfrac{f(x+h) - f(x)}{h}$, $h \neq 0$ for $f(x) = 2x^2 + 4$. Give answer in simplified form.

A. $4hx + 2h$ B. $4x + 2h^2$ C. $4x$

D. $4x + h$ E. $4x + 2h$

Test form RC Date _________ Name ________________________
Course ______ Sec ______ Student Number ________________

Chapter R

R-1. Let $X = \left\{-9, -\frac{2}{7}, 0, \sqrt{6}\right\}$. List all the elements of X that are rational numbers.

A. $-9, -\frac{2}{7}$

B. $-9, -\frac{2}{7}, 0$

C. $-9, -\frac{2}{7}, \sqrt{6}$

D. -9

E. $\sqrt{6}$

R-2. Indicate which algebraic property of real numbers is used to justify the statement $8x + 7y = 7y + 8x$.

A. Commutative property of multiplication

B. Commutative property of addition

C. Associative property of multiplication

D. Associative property of addition

E. Distributive property

R-3. Compute the expression $|7.14 - 8.19| \div 3.14$ to four decimal places.

A. -1.05

B. -0.3344

C. 0.3344

D. 1.05

E. 4.5317

R-4. Evaluate $\sqrt[8]{(-5)^8}$

A. 5

B. -5

C. 1.4953

D. -1.4953

E. 2.2361

R-5. Use the rules of exponents to write the algebraic expression in a form that involves each variable at most once to a positive exponent.

$$\frac{4^3 x^6 y^6}{2^3 x^9 y^3}$$

A. $\frac{8y^3}{x^3}$ B. $8x^{2/5}y^{7/4}$ C. $\frac{y^3}{8x^3}$

D. $8x^3y^3$ E. $\frac{8x^3}{y^3}$

R-6. Simplify $\sqrt[6]{7x^6y^8}$ by removing as many factors as possible from under the radical.

A. $|xy|\sqrt[6]{7y^2}$ B. $xy\sqrt[6]{7y^2}$ C. $xy\sqrt{7y}$

D. $-xy\sqrt{7y}$ E. $7xy\sqrt{y}$

R-7. Use a calculator to compute $5.13^{-1.3}$. Express the result to four decimal places.

A. 0.1194 B. 0.1121 C. 0.1125

D. 8.3782 E. 8.3791

R-8. Give the degree of the polynomial. $11x^6 + 4x^3 - 7$

A. 3 B. 11 C. 17 D. 6 E. 18

R-9. Perform the operations of addition and subtraction of polynomials.
$(5x^2 + x + 7) + (2x^2 + 7x + 2) - (3x^2 + 5x + 7)$

A. $10x^2 + 13x + 16$ B. $4x^2 - 13x - 16$

C. $4x^2 + 3x + 2$ D. $4x^2 + 13x + 16$

E. $4x^2 - 3x - 2$

R-10. Multiply the polynomials. $(4x^3+3)(2x+5)$

A. $8x^2+26x+15$ B. $8x^2+15$

C. $8x^4+15$ D. $8x^4+20x^3+6x+15$

E. $8x^4+26x^2+15$

R-11. Divide x^2-4x+3 by $x+2$.

A. $x-6+\dfrac{9}{x+2}$ B. $x-6+\dfrac{15}{x+2}$

C. $x-6+\dfrac{-15}{x+2}$ D. $x-2+\dfrac{7}{x+2}$

E. $x-2+\dfrac{-7}{x+2}$

R-12. Factor $x^2+3x-28$.

A. $(x-7)(x+4)$ B. $(x+7)(x-4)$

C. $(x-7)(x-4)$ D. $(x-28)(x-1)$

E. $(x-28)(x+1)$

R-13. The expression is the difference of two squares. Factor $49x^2-(3y+1)^2$.

A. $(7x+3y+1)^2$ B. $(7x-3y-1)(7x+3y+1)$

C. $(7x-3y-1)(7x-3y-1)$ D. $(49x-3y+1)(49x+3y+1)$

E. $(49-3y+1)(49+3y+1)$

R-14. Factor the perfect square $25x^2 - 20x + 4$.

A. $(5x+1)^2$ B. $(5x-1)^2$ C. $(5x+2)^2$

D. $(5x-2)^2$ E. $(25x-4)^2$

R-15. Simplify the rational expression. $\frac{(x^2+10x-24)}{(x^2+16x-36)}$

A. $\frac{x-12}{x-18}$ B. $\frac{x+12}{x+18}$ C. $\frac{x-6}{x-9}$

D. $\frac{x+6}{x+9}$ E. $\frac{x^2+4x-12}{x^2+7x-18}$

R-16. Perform the multiplication. $\frac{(7x+3)^2}{49} \cdot \frac{14}{7x+3}$

A. $\frac{2(7x+3)}{7}$ B. $(7x+3)$ C. $2(x+3)$

D. $49(7x+3)$ E. $2(7x+3)$

R-17. Perform the indicated operations. $\frac{x^2-x-20}{x^2+3x-4} \cdot \frac{x^2+4x-5}{x^2-25}$

A. 1 B. $\frac{x-5}{x+5}$ C. $\frac{x-4}{x+4}$

D. $\frac{x+4}{x-4}$ E. $\frac{x+5}{x-5}$

R-18. Perform the indicated operation. $\frac{6}{x-1}+\frac{9}{3x+1}$

A. $\frac{15}{4x}$

B. $\frac{1}{(x-1)(3x+1)}$

C. $\frac{27x-3}{(x-1)(3x+1)}$

D. $\frac{27x+3}{(x-1)(3x+1)}$

E. $\frac{-1}{(x-1)(3x+1)}$

R-19. Solve the equation for x. $4x+2=6$

A. -1 B. 1 C. 2 D. -2 E. 4

R-20. Solve the equation for x. $6(x+3)=12(2x+1)$

A. $\frac{1}{3}$ B. 3 C. -3 D. $-\frac{1}{3}$ E. 0

R-21. Solve the equation for x, if a solution exists. $\frac{x}{x-4}-\frac{4}{x-4}=4$

A. 4

B. 0

C. 1

D. -4

E. No solution

R-22. If $f(x)=5x^2+4$, find $f(2)$.

A. 24 B. 0 C. 16 D. 9 E. 1

R-23. Find the domain of $f(x) = \dfrac{1}{\sqrt{x-12}}$.

A. $(12, \infty)$ B. $[12, \infty)$ C. $(-\infty, 12)$

D. $(-\infty, 12]$ E. $(-\infty, -12)$

R-24. Compute $f(-3.7)$ for $f(x) = 5.01x^2 + 2.31x + 4.17$ to four decimal places.

A. 64.1702 B. 64.1684 C. 64.2099

D. 64.0912 E. 64.0977

R-25. Determine the difference quotient $\dfrac{f(x+h) - f(x)}{h}$, $h \neq 0$ for $f(x) = 2x^2 + 5$.

Give answer in simplified form.

A. $4hx + 2h$ B. $4x + 2h^2$ C. $4x$

D. $4x + h$ E. $4x + 2h$

Test form RD Date __________ Name ____________________________
Course _______ Sec _______ Student Number ___________________

Chapter R

R-1. Let $X = \left\{-11, -\frac{3}{7}, 0, \sqrt{5}\right\}$. List all the elements of X that are rational numbers.

R-2. Indicate which algebraic property of real numbers is used to justify the statement $7x + 7y = 7y + 7x$.

R-3. Compute the expression $|6.42 - 7.14| \div 5.17$ to four decimal places.

R-4. Evaluate $\sqrt[8]{(-6)^8}$

R-5. Use the rules of exponents to write the algebraic expression in a form that involves each variable at most once to a positive exponent.

$$\frac{9^3 x^4 y^5}{3^3 x^8 y^3}$$

R-6. Simplify $\sqrt[6]{3x^6 y^{10}}$ by removing as many factors as possible from under the radical.

R-7. Use a calculator to compute $3.18^{-2.8}$. Express the result to four decimal places.

R-8. Give the degree of the polynomial. $14x^8 + 5x^5 - 7$

R-9. Perform the operations of addition and subtraction of polynomials.
$(9x^2 + 4x + 3) + (x^2 + 5x + 7) - (3x^2 + 4x + 5)$

R-10. Multiply the polynomials. $(5x^3 + 4)(7x + 3)$

R-11. Divide $x^2 + 10x + 9$ by $x + 2$.

R-12. Factor $x^2 + 3x - 40$.

R-13. The expression is the difference of two squares. Factor $81x^2 - (5y + 3)^2$.

R-14. Factor the perfect square $4x^2 - 28x + 49$.

R-15. Simplify the rational expression. $\dfrac{(x^2 + 4x - 21)}{(x^2 + x - 12)}$

R-16. Perform the multiplication. $\dfrac{(5x+9)^2}{81} \cdot \dfrac{18}{5x+9}$

R-17. Perform the indicated operations. $\dfrac{x^2 + 2x - 48}{x^2 + 7x - 8} \cdot \dfrac{x^2 + 5x - 6}{x^2 - 36}$

R-18. Perform the indicated operation. $\dfrac{7}{x-1} + \dfrac{10}{3x+1}$

R-19. Solve the equation for x. $7x + 2 = 16$

R-20. Solve the equation for x. $10(x+3) = 20(2x+1)$

R-21. Solve the equation for x, if a solution exists. $\dfrac{x}{x-5} - \dfrac{5}{x-5} = 4$

R-22. If $f(x) = 6x^2 + 5$, find $f(3)$.

R-23. Find the domain of $f(x) = \dfrac{1}{\sqrt{x-5}}$.

R-24. Compute $f(-1.7)$ for $f(x) = 7.01x^2 + 4.02x + 9.31$ to four decimal places.

R-25. Determine the difference quotient $\dfrac{f(x+h) - f(x)}{h}$, $h \neq 0$ for $f(x) = 5x^2 + 7$.
Give answer in simplified form.

Test form RE Date _________ Name ___________________________
Course ______ Sec ______ Student Number _________________

Chapter R

R-1. Let $X = \left\{-15, -\frac{2}{5}, 0, \sqrt{8}\right\}$. List all the elements of X that are rational numbers.

R-2. Indicate which algebraic property of real numbers is used to justify the statement $6x + 8y = 8y + 6x$.

R-3. Compute the expression $|4.17 - 8.15| \div 6.97$ to four decimal places.

R-4. Evaluate $\sqrt[4]{(-3)^4}$

R-5. Use the rules of exponents to write the algebraic expression in a form that involves each variable at most once to a positive exponent.

$$\frac{9^3 x^6 y^7}{3^3 x^8 y^3}$$

R-6. Simplify $\sqrt[4]{5x^4 y^6}$ by removing as many factors as possible from under the radical.

R-7. Use a calculator to compute $5.91^{-3.7}$. Express the result to four decimal places.

R-8. Give the degree of the polynomial. $10x^8 + 4x^3 - 6$

R-9. Perform the operations of addition and subtraction of polynomials.
$(7x^2 + 7x + 9) + (2x^2 + x + 3) - (6x^2 + 5x + 8)$

R-10. Multiply the polynomials. $(2x^3 + 9)(4x + 10)$

R-11. Divide $x^2 + 12x + 1$ by $x + 2$.

R-12. Factor $x^2 + x - 42$.

R-13. The expression is the difference of two squares. Factor $64x^2 - (7y - 4)^2$.

R-14. Factor the perfect square $9x^2 - 42x + 49$.

R-15. Simplify the rational expression. $\dfrac{(x^2 + 6x - 16)}{(x^2 + 2x - 8)}$

R-16. Perform the multiplication. $\dfrac{(4x + 11)^2}{9} \cdot \dfrac{6}{4x + 11}$

R-17. Perform the indicated operations. $\dfrac{x^2 + x - 20}{x^2 + 4x - 5} \cdot \dfrac{x^2 + 3x - 4}{x^2 - 16}$

R-18. Perform the indicated operation. $\dfrac{7}{x - 1} + \dfrac{4}{3x + 1}$

R-19. Solve the equation for x. $6x + 3 = 15$

R-20. Solve the equation for x. $2(x + 3) = 4(2x + 1)$

R-21. Solve the equation for x, if a solution exists. $\dfrac{x}{x - 6} - \dfrac{6}{x - 6} = 4$

R-22. If $f(x) = 2x^2 + 9$, find $f(4)$.

R-23. Find the domain of $f(x) = \dfrac{1}{\sqrt{x - 13}}$.

R-24. Compute $f(-1.9)$ for $f(x) = 6.49x^2 + 5.92x + 8.32$ to four decimal places.

R-25. Determine the difference quotient $\dfrac{f(x + h) - f(x)}{h}$, $h \neq 0$ for $f(x) = 6x^2 + 9$.

Give answer in simplified form.

Test form RF Date ________ Name ____________________
Course ______ Sec ______ Student Number ________________

Chapter R

R-1. Let $X = \left\{-13, -\frac{4}{7}, 0, \sqrt{3}\right\}$. List all the elements of X that are rational numbers.

R-2. Indicate which algebraic property of real numbers is used to justify the statement $5x + 9y = 9y + 5x$.

R-3. Compute the expression $|2.97 - 9.41| \div 3.62$ to four decimal places.

R-4. Evaluate $\sqrt[4]{(-9)^4}$

R-5. Use the rules of exponents to write the algebraic expression in a form that involves each variable at most once to a positive exponent.

$$\frac{9^3 x^8 y^{10}}{3^3 x^{12} y^7}$$

R-6. Simplify $\sqrt[8]{11x^8 y^{10}}$ by removing as many factors as possible from under the radical.

R-7. Use a calculator to compute $6.14^{-2.4}$. Express the result to four decimal places.

R-8. Give the degree of the polynomial. $9x^7 + 3x^5 - 8$

R-9. Perform the operations of addition and subtraction of polynomials.
$(9x^2 + 2x + 7) + (3x^2 + 4x + 1) - (8x^2 + 3x + 4)$

R-10. Multiply the polynomials. $(5x^3 + 7)(6x + 8)$

R-11. Divide $x^2 + 8x + 9$ by $x + 2$.

R-12. Factor $x^2 + 6x - 16$.

R-13. The expression is the difference of two squares. Factor $16x^2 - (9y + 1)^2$.

R-14. Factor the perfect square $16x^2 - 40x + 25$.

R-15. Simplify the rational expression. $\dfrac{(x^2+6x-27)}{(x^2+2x-15)}$

R-16. Perform the multiplication. $\dfrac{(6x+11)^2}{121} \cdot \dfrac{22}{6x+11}$

R-17. Perform the indicated operations. $\dfrac{x^2+8x-20}{x^2+9x-10} \cdot \dfrac{x^2+x-2}{x^2-4}$

R-18. Perform the indicated operation. $\dfrac{3}{x-1} + \dfrac{2}{3x+1}$

R-19. Solve the equation for x. $7x+1=15$

R-20. Solve the equation for x. $8(x+3) = 16(2x+1)$

R-21. Solve the equation for x, if a solution exists. $\dfrac{x}{x-7} - \dfrac{7}{x-7} = 4$

R-22. If $f(x) = 3x^2 + 5$, find $f(2)$.

R-23. Find the domain of $f(x) = \dfrac{1}{\sqrt{x-6}}$.

R-24. Compute $f(-2.8)$ for $f(x) = 3.11x^2 + 3.99x + 4.12$ to four decimal places.

R-25. Determine the difference quotient $\dfrac{f(x+h) - f(x)}{h}$, $h \neq 0$ for $f(x) = 9x^2 + 7$.
Give answer in simplified form.

Test form 1A　　Date ________　　Name ______________________
Course ______ Sec ______　　Student Number ________________

Chapter 1

1-1. Determine which of the given points lie on the graph of $y = 3x^2 + 4x + 6$. (1, 6), (0, 6), (1, 13)

A. (1, 6) only
B. (0, 6) only
C. (1, 13) only
D. (0, 6) and (1, 13)
E. (1, 6) and (1, 13)

1-2. Find the x- and y-intercepts of the graph of $y = 4x - 8$ by letting y and then x be zero.

A. x-intercept (8, 0), y-intercept (0, 4)

B. x-intercept (2, 0), y-intercept (0, 8)

C. x-intercept (2, 0), y-intercept (0, –8)

D. x-intercept (–8, 0), y-intercept (0, 4)

E. x-intercept (–2, 0), y-intercept (0, 8)

1-3. The points at which the line crosses the x and y axes are the integers 2 and –4, respectively. Find the slope of the line.

A. 6　　B. 2　　C. –3　　D. 3　　E. –6

1-4. Use the point-slope form, $y - y_1 = m(x - x_1)$, with $m = \frac{y_2 - y_1}{x_2 - x_1}$ to find the equation of the line through the points (2, 5) and (3, 9).

A. $y - 5 = 4(x - 2)$
B. $y - 2 = 4(x - 5)$
C. $y - 5 = \frac{1}{4}(x - 2)$
D. $y - 2 = \frac{1}{4}(x - 5)$
E. $y = 4x$

1-5. Find the equation of the horizontal line through (4, 7).

A. $x = 4$ B. $y = 7$ C. $4x = 7y$

D. $4y = 7x$ E. $x - 4 = y - 7$

1-6. Find the equation of the vertical line through (4, 7).

A. $x = 4$ B. $y = 7$ C. $4x = 7y$

D. $4y = 7x$ E. $x - 4 = y - 7$

1-7. Find an equation of the line *M* through the point (5, 7) that is parallel to $9x - 3y = 6$. Give answer in point-slope form.

A. $y - 7 = 3(x - 5)$ B. $y + 7 = 3(x + 5)$ C. $y - 7 = \frac{1}{3}(x - 5)$

D. $y - 7 = -\frac{1}{3}(x - 5)$ E. $y + 7 = -\frac{1}{3}(x + 5)$

1-8. Find an equation of the line *M* through the point (5, 7) that is perpendicular to $9x - 3y = 6$. Give answer in point-slope form.

A. $y - 7 = 3(x - 5)$ B. $y + 7 = 3(x + 5)$ C. $y - 7 = \frac{1}{3}(x - 5)$

D. $y - 7 = -\frac{1}{3}(x - 5)$ E. $y + 7 = -\frac{1}{3}(x + 5)$

1-9. Use a calculator to determine the least squares line through (1, 0), (2, 3), (3, 8), and (4, 13).

A. $x = 4.4y$ B. $x = -4.4y + 5$ C. $x = 4.4y - 5$

D. $y = 4.4 - 5x$ E. $y = 4.4x - 5$

1-10. Use a calculator to find the value of the linear function $f(x) = 7.21x - 6.14$ at $x = -3.12$.

A. −28.64 B. −16.36 C. 16.36 D. 28.64 E. −3.34

1-11. Determine which of these tables is for a linear function.

A.

x	y
0	3.1
1	6.2
2	9.3
3	12.4

B.

x	y
0	2.1
1	5.5
2	8.6
3	10.5

C.

x	y
0	2
1	4
2	9
3	20

D.

x	y
0	1
1	3
2	12
3	18

E.

x	y
0	3
1	7
2	21
3	26

1-12. A refrigerator costs $620 and there is a cost of $42 per year to run it. Construct a function that describes the cost of owning and running the refrigerator.

A. $y = 620x - 42$ B. $y = 620x/42$ C. $y = 42x + 620$

D. $y = 620x + 42$ E. $y = (620x)(42)$

1-13. Find the vertex of $y = 2x^2 + 4x - 8$.

A. (1, 6) B. (−1, −10) C. (1, −2) D. (0, −8) E. (0, 8)

1-14. Find the axis of symmetry of the graph $y = 2x^2 + 4x - 8$.

A. $y = 10$ B. $y = -10$ C. $x = -1$ D. $x = 1$ E. $x = 0$

1-15. Find the y-intercept of the graph of $y = 4x^2 + 2x - 8$.

A. (0, 8) B. (0, −8) C. (8, 0) D. (−8, 0) E. (0, 0)

1-16. Find the x-intercepts of the parabola by factoring the equation $y = x^2 - 7x + 12$.

A. (−12, 0) and (12, 0) B. (0, −12) and (0, 12) C. (−3, 0) and (−4, 0)

D. (3, 0) and (4, 0) E. (0, 3) and (0, 4)

1-17. Find the x-intercepts of the graph of $y = 3x^2 - 5x + 1$.

A. (0.2321, 0) and (1.4338, 0)
B. (0.2324, 0) and (1.4338, 0)
C. (0.2324, 0) and (1.4343, 0)
D. (0.2330, 0) and (1.4343, 0)
E. (0.2330, 0) and (1.4350, 0)

1-18. Find p so that (2, 4) lies on the graph of $y = x^2 + 4px - 40$.

A. 5 B. 0 C. 4 D. 8 E. 3

1-19. If an object is propelled vertically from the ground level with a velocity v, its motion in meters is described by the quadratic function $s(t) = -16t^2 + vt$. A particular object which is propelled reaches its maximum height in 16 seconds. What is the object's initial velocity?

A. 256 m/s B. 512 m/s C. 128 m/s D. 16 m/s E. 32 m/s

1-20. Find the points of intersection of $y = 0.3x^2 - x + 5$ and $y = -2x^2 + 3x + 7$.

A. (–0.4055, 4.2352) and (2.1446, 5.4548)

B. (–0.4055, 5.4548) and (2.1446, 4.2352)

C. (–0.4051, 5.4540) and (2.1441, 4.2351)

D. (–0.4050, 4.2360) and (2.1448, 5.4552)

E. (–0.4059, 4.2351) and (2.1440, 5.4540)

1-21. Use the distance formula to compute the distance between (1, 5) and (3, –2) to four decimals.

A. 9.0000 B. 7.2801 C. 5.0000 D. 3.6056 E. 52.9999

1-22. Use the midpoint formula to find the midpoint of the line segment joining (2, 4) and (–2, 0).

A. (1, 1) B. (0, 0) C. (2, 2) D. (0, 2) E. (1, 2)

1-23. Find the center and radius of the circle having the equation $(x+3)^2+(y-5)^2=4$.

A. Center: (3, –5), Radius: 2
B. Center: (–3, 5), Radius: 2
C. Center: (–3, 5), Radius: 4
D. Center: (3, –5), Radius: 4
E. Center: (–3, 5), Radius: 16

1-24. Determine the equation of the circle having center (–5, 6) and radius 3.

A. $(x+5)^2+(y-6)^2=3$
B. $(x+5)^2+(y-6)^2=9$
C. $(x-5)^2+(y+6)^2=3$
D. $(x-5)^2+(y+6)^2=9$
E. $(x+5)^2+(y-6)^2=\sqrt{3}$

1-25. Find the center and radius of the circle having the equation $x^2-4x+y^2+6y=17$.

A. Center: (–2, –3), Radius: 4
B. Center: (2, 3), Radius: 4
C. Center: (2, 3), Radius: 2
D. Center: (2, 3), Radius: $\sqrt{30}$
E. Center: (–2, –3), Radius: 2

Test form 1B Date ________ Name ________________________
Course ______ Sec ______ Student Number ______________

Chapter 1

1. Determine which of the given points lie on the graph of $y = 2x^2 + x + 6$. (1, 9), (0, 9), (0, 6)

   A. (1, 9) only B. (0, 9) only

   C. (0, 6) only D. (1, 9) and (1, 6)

   E. (0, 6) and (1, 9)

2. Find the x- and y-intercepts of the graph of $y = 4x - 12$ by letting y and then x be zero.

   A. x-intercept (12, 0), y-intercept (0, 4)

   B. x-intercept (3, 0), y-intercept (0, 12)

   C. x-intercept (3, 0), y-intercept (0, –12)

   D. x-intercept (–12, 0), y-intercept (0, 4)

   E. x-intercept (–3, 0), y-intercept (0, 12)

3. The points at which the line crosses the x and y axes are the integers 2 and –6, respectively. Find the slope of the line.

   A. 6 B. 2 C. –3 D. 3 E. –6

4. Use the point-slope form, $y - y_1 = m(x - x_1)$, with $m = \frac{y_2 - y_1}{x_2 - x_1}$ to find the equation of the line through the points (3, 6) and (4, 10).

   A. $y - 3 = 4(x - 6)$ B. $y - 6 = 4(x - 3)$ C. $y - 3 = \frac{1}{4}(x - 6)$

   D. $y - 3 = \frac{1}{4}(x - 6)$ E. $y = 4x$

5. Find the equation of the horizontal line through (3, 6).

A. $y = 6$ B. $x - 3$ C. $3x = 6y$

D. $3y = 6x$ E. $x - 6 = y - 3$

6. Find the equation of the vertical line through (3, 7).

A. $y = 6$ B. $x = 3$ C. $3x = 6y$

D. $3y = 6x$ E. $x - 6 = y - 3$

7. Find an equation of the line *M* through the point (8, 2) that is parallel to $9x - 3y = 6$. Give answer in point-slope form.

A. $y + 2 = 3(x + 8)$ B. $y - 2 = 3(x - 8)$ C. $y - 2 = \frac{1}{3}(x - 8)$

D. $y + 2 = -\frac{1}{3}(x + 8)$ E. $y - 2 = -\frac{1}{3}(x - 8)$

8. Find an equation of the line *M* through the point (8, 2) that is perpendicular to $9x - 3y = 6$. Give answer in point-slope form.

A. $y + 2 = 3(x + 8)$ B. $y - 7 = 3(x - 2)$ C. $y - 2 = \frac{1}{3}(x - 8)$

D. $y + 2 = -\frac{1}{3}(x + 8)$ E. $y - 2 = -\frac{1}{3}(x - 8)$

9. Use a calculator to determine the least squares line through (1, 1), (2, 4), (3, 9), and (4, 14).

A. $x = 4.4y$ B. $x = -4.4y + 4$ C. $y = 4.4x - 4$

D. $y = 4.4 - 4x$ E. $x = 4.4y - 4$

10. Use a calculator to find the value of the linear function $f(x) = 7.21x - 6.14$ at $x = -4.13$.

A. −23.64 B. 35.92 C. 23.64 D. −35.92 E. −4.13

11. Determine which of these tables is for a linear function.

A.

x	y
0	4.2
1	5.3
2	6.4
3	7.5

B.

x	y
0	2.1
1	5.9
2	8.3
3	10.2

C.

x	y
0	3
1	5
2	8
3	16

D.

x	y
0	1
1	3
2	9
3	13

E.

x	y
0	2
1	8
2	32
3	81

12. A refrigerator costs \$730 and there is a cost of \$52 per year to run it. Construct a function that describes the cost of owning and running the refrigerator.

A. $y = 730x - 52$ B. $y = 730x/52$ C. $y = 52x + 730$

D. $y = 730x + 52$ E. $y = (730x)(52)$

13. Find the vertex of $y = 2x^2 + 6x - 8$.

A. (1.5, 12.5)B. (1, 0) C. (0, −8) D. (−1.5, −12.5) E. (0, 8)

14. Find the axis of symmetry of the graph $y = 2x^2 + 6x - 8$.

A. $y = 12.5$ B. $y = -12.5$ C. $x = -1.5$ D. $x = 0$ E. $x = 1.5$

15. Find the y-intercept of the graph of $y = 2x^2 + 6x - 8$.

A. (0, 8) B. (8, 0) C. (0, −8) D. (−8, 0) E. (0, 0)

16. Find the x-intercepts of the parabola by factoring the equation $y = x^2 - 8x + 12$.

A. (2, 0) and (6, 0) B. (0, −12) and (0, 12) C. (−2, 0) and (−6,
0)

A. (−12, 0) and (12, 0) E. (0, 2) and (0, 6)

17. Find the x-intercepts of the graph of $y = 7x^2 - 5x - 1$.

A. (–0.1604, 0) and (0, 0.8762) B. (–0.1610, 0) and (0, 0.8762)

C. (–0.1610, 0) and (0, 0.8772) D. (–0.1620, 0) and (0, 0.8772)

E. (–0.1629, 0) and (0, 0.8772)

18. Find p so that (1, 3) lies on the graph of $y = x^2 + 4px - 10$.

A. 5 B. 0 C. 4 D. 8 E. 3

19. If an object is propelled vertically from the ground level with a velocity v, its motion in meters is described by the quadratic function $s(t) = -16t^2 + vt$. A particular object which is propelled reaches its maximum height in 8 seconds. What is the object's initial velocity?

A. 256 m/s B. 512 m/s C. 128 m/s D. 16 m/s E. 32 m/s

20. Find the points of intersection of $y = 0.3x^2 - x + 4$ and $y = -2x^2 + 3x + 6$.

A. (–0.4055, 3.2352) and (2.1446, 4.4548)

B. (–0.4055, 4.4548) and (2.1446, 3.2352)

C. (–0.4051, 4.4540) and (2.1441, 3.2351)

D. (–0.4050, 3.2360) and (2.1448, 4.4552)

E. (–0.4059, 3.2351) and (2.1440, 4.4540)

21. Use the distance formula to compute the distance between (2, 6) and (3, –2) to four decimals.

A. 9.0000 B. 8.0623 C. 7.2801 D. 3.6056 E. 52.9999

22. Use the midpoint formula to find the midpoint of the line segment joining (2, 8) and (–2, 0).

A. (1, 1) B. (0, 0) C. (2, 4) D. (1, 4) E. (0, 4)

23. Find the center and radius of the circle having the equation $(x+2)^2 + (y-8)^2 = 4$.

A. Center: (2, –8), Radius: 2 B. Center: (–2, 8), Radius: 4

C. Center: (–2, 8), Radius: 2 D. Center: (2, –8), Radius: 4

E. Center: (–2, 8), Radius: 16

24. Determine the equation of the circle having center (–2, 9) and radius 3.

A. $(x+2)^2 + (y-9)^2 = 3$ B. $(x-2)^2 + (y+9)^2 = 9$

C. $(x-2)^2 + (y+9)^2 = 3$ D. $(x+2)^2 + (y-9)^2 = 9$

E. $(x+2)^2 + (y-9)^2 = \sqrt{3}$

25. Find the center and radius of the circle having the equation $x^2 - 4x + y^2 - 6y = 22$.

A. Center: (–2, –3), Radius: 9 B. Center: (2, 3), Radius: 9

C. Center: (2, 3), Radius: 3 D. Center: (2, 3), Radius: $\sqrt{35}$

E. Center: (–2, –3), Radius: 3

Test form 1C Date _________ Name ________________________

Course ______ Sec ______ Student Number _______________

Chapter 1

1. Determine which of the given points lie on the graph of $y = 4x^2 - 4x + 6$. (1, 6), (0, 6), (1, 14)

   A. (1, 6) only   B. (0, 6) and (1, 6)

   C. (1, 14) only   D. (0, 6) and (1, 14)

   E. (0, 6) only

2. Find the x- and y-intercepts of the graph of $y = 2x - 8$ by letting y and then x be zero.

   A. x-intercept (8, 0), y-intercept (0, 2)

   B. x-intercept (4, 0), y-intercept (0, 8)

   C. x-intercept (–8, 0), y-intercept (0, 2)

   D. x-intercept (–4, 0), y-intercept (0, 8)

   E. x-intercept (4, 0), y-intercept (0, –8)

3. The points at which the line crosses the x and y axes are the integers 4 and –8, respectively. Find the slope of the line.

   A. 6   B. 2   C. –3   D. 3   E. –6

4. Use the point-slope form, $y - y_1 = m(x - x_1)$, with $m = \frac{y_2 - y_1}{x_2 - x_1}$ to find the equation of the line through the points (4, 7) and (5, 11).

   A. $y - 7 = 4(x - 4)$ B. $y - 4 = 4(x - 7)$ C. $y - 7 = \frac{1}{4}(x - 4)$

   D. $y - 4 = \frac{1}{4}(x - 7)$ E. $y = 4x$

5. Find the equation of the horizontal line through (5, 8).

A. $x = 5$ B. $5x = 8y$ C. $y = 8$

D. $5y = 8x$ E. $x - 5 = y - 8$

6. Find the equation of the vertical line through (5, 8).

A. $x = 5$ B. $5x = 8y$ C. $y = 8$

D. $5y = 8x$ E. $x - 5 = y - 8$

7. Find an equation of the line M through the point (9, 2) that is parallel to $9x - 3y = 6$. Give answer in point-slope form.

A. $y - 2 = \frac{1}{3}(x - 9)$ B. $y - 2 = -\frac{1}{3}(x - 9)$ C. $y + 2 = -\frac{1}{3}(x + 9)$

D. $y - 2 = 3(x - 9)$ E. $y + 2 = 3(x + 9)$

8. Find an equation of the line M through the point (9, 2) that is perpendicular to $9x - 3y = 6$. Give answer in point-slope form.

A. $y - 2 = \frac{1}{3}(x - 9)$ B. $y - 2 = -\frac{1}{3}(x - 9)$ C. $y + 2 = -\frac{1}{3}(x + 9)$

D. $y - 2 = 3(x - 9)$ E. $y + 2 = 3(x + 9)$

9. Use a calculator to determine the least squares line through (1, 3), (2, 6), (3, 11), and (4, 16).

A. $x = 4.4y$ B. $x = -4.4y + 2$ C. $x = 4.4y - 2$

D. $y = 4.4 - x$ E. $y = 4.4x - 2$

10. Use a calculator to find the value of the linear function $f(x) = 4.81x - 9.03$ at $x = -3.07$.

A. −5.74 B. 23.80 C. 5.74 D. −23.80 E. −3.07

11. Determine which of these tables is for a linear function.

A.

x	y
0	2.2
1	3.3
2	4.4
3	5.5

B.

x	y
0	2.2
1	5.5
2	8.3
3	10.1

C.

x	y
0	2
1	5
2	8
3	19

D.

x	y
0	1
1	4
2	6
3	16

E.

x	y
0	3
1	5
2	24
3	90

12. A refrigerator costs \$640 and there is a cost of \$84 per year to run it. Construct a function that describes the cost of owning and running the refrigerator.

A. $y = 640x - 84$ B. $y = 640x/84$ C. $y = 84x + 640$

D. $y = 640x + 84$ E. $y = (640x)(84)$

13. Find the vertex of $y = 2x^2 - 8x - 4$.

A. $(2, -12)$ B. $(-2, 12)$ C. $(1, -2)$ D. $(0, -4)$ E. $(0, 4)$

14. Find the axis of symmetry of the graph $y = 2x^2 - 8x - 84$

A. $y = 12$ B. $y = -12$ C. $x = -2$ D. $x = 2$ E. $x = 0$

15. Find the y-intercept of the graph of $y = 2x^2 - 8x - 4$.

A. $(0, 4)$ B. $(0, -4)$ C. $(4, 0)$ D. $(-4, 0)$ E. $(0, 0)$

16. Find the x-intercepts of the parabola by factoring the equation $y = x^2 + 7x + 12$.

A. $(-12, 0)$ and $(12, 0)$ B. $(0, -12)$ and $(0, 12)$ C. $(-3, 0)$ and $(-4, 0)$

D. $(3, 0)$ and $(4, 0)$ E. $(0, 3)$ and $(0, 4)$

17. Find the x-intercepts of the graph of $y = 3x^2 - 2x - 6$.

A. (−1.1196, 0) and (1.7863, 0) B. (−1.1120, 0) and (1.7865, 0)

C. (−1.1134, 0) and (1.7865, 0) C. (−1.1123, 0) and (1.7872, 0)

E. (−1.1121, 0) and (1.7883, 0)

18. Find p so that (1, 3) lies on the graph of $y = x^2 + 4px - 14$.

A. 5 B. 0 C. 4 D. 8 E. 3

19. If an object is propelled vertically from the ground level with a velocity v, its motion in meters is described by the quadratic function $s(t) = -16t^2 + vt$. A particular object which is propelled reaches its maximum height in 2 seconds. What is the object's initial velocity?

A. 256 m/s B. 512 m/s C. 128 m/s D. 16 m/s E. 64 m/s

20. Find the points of intersection of $y = 0.3x^2 - x + 6$ and $y = -2x^2 + 3x + 8$.

A. (−0.4055, 5.2352) and (2.1446, 6.4548)

B. (−0.4055, 6.4548) and (2.1446, 5.2352)

C. (−0.4051, 6.4540) and (2.1441, 5.2351)

D. (−0.4050, 5.2360) and (2.1448, 6.4552)

E. (−0.4059, 5.2351) and (2.1440, 6.4540)

21. Use the distance formula to compute the distance between (1, −3) and (3, 4) to four decimals.

A. 9.0000 B. 5.0000 C. 3.6056 D. 7.2801 E. 52.9999

22. Use the midpoint formula to find the midpoint of the line segment joining (2, 6) and (–2, 0).

A. (1, 1) B. (0, 0) C. (0, 3) D. (2, 3) E. (1, 3)

23. Find the center and radius of the circle having the equation $(x+1)^2 + (y-4)^2 = 4$.

A. Center: (–1, 4), Radius: 2
B. Center: (1, –4), Radius: 2
C. Center: (–1, 4), Radius: 4
D. Center: (1, –4), Radius: 4
E. Center: (–3, 5), Radius: 16

24. Determine the equation of the circle having center (–3, 7) and radius 3.

A. $(x+3)^2 + (y-7)^2 = 3$
B. $(x-3)^2 + (y+7)^2 = 3$
C. $(x+3)^2 + (y-7)^2 = 9$
D. $(x-3)^2 + (y+7)^2 = 9$
E. $(x+3)^2 + (y-7)^2 = \sqrt{3}$

25. Find the center and radius of the circle having the equation $x^2 + 6x + y^2 + 4y = 17$.

A. Center: (–3, –2), Radius: 4
B. Center: (3, 2), Radius: 4
C. Center: (3, 2), Radius: 2
D. Center: (–3, –2), Radius: $\sqrt{30}$
E. Center: (–3, –2), Radius: 2

Test form 1D Date _________ Name _________________________
Course ______ Sec ______ Student Number ________________

Chapter 1

1. Determine which of the given points lie on the graph of $y = 4x^2 + 5x + 6$. (1, 6), (0, 6), (1, 15)

2. Find the x- and y-intercepts of the graph of $y = 6x - 12$ by letting y and then x be zero.

3. The points at which the line crosses the x and y axes are the integers 3 and –6, respectively. Find the slope of the line.

4. Use the point-slope form, $y - y_1 = m(x - x_1)$, with $m = \dfrac{y_2 - y_1}{x_2 - x_1}$ to find the equation of the line through the points (4, 9) and (6, 2).

5. Find the equation of the horizontal line through (2, 1).

6. Find the equation of the vertical line through (2, 1).

7. Find an equation of the line M through the point (2, 1) that is parallel to $12x - 3y = 6$. Give answer in point-slope form.

8. Find an equation of the line M through the point (5, 7) that is perpendicular to $12x - 3y = 6$. Give answer in point-slope form.

9. Use a calculator to determine the least squares line through (1, 2), (2, 5), (3, 9), and (4, 11).

10. Use a calculator to find the value of the linear function $f(x) = 5.24x - 8.17$ at $x = -3.36$

11. Complete the table for a linear function.

x	y
0	5
1	10
2	15
3	20
4	__

12. A refrigerator costs \$935 and there is a cost of \$67 per year to run it. Construct a function that describes the cost of owning and running the refrigerator.

13. Find the vertex of $y = 3x^2 + 6x - 7$.

14. Find the axis of symmetry of the graph $y = 3x^2 + 6x - 7$.

15. Find the y-intercept of the graph of $y = 3x^2 + 6x - 7$.

16. Find the x-intercepts of the parabola by factoring the equation $y = x^2 - 8x + 15$.

17. Find the x-intercepts of the graph of $y = 5x^2 - 6x - 9$.

18. Find p so that (1, 4) lies on the graph of $y = x^2 + 4px - 17$.

19. If an object is propelled vertically from the ground level with a velocity v, its motion in meters is described by the quadratic function $s(t) = -16t^2 + vt$. A particular object which is propelled reaches its maximum height in 10 seconds. What is the object's initial velocity?

20. Find the points of intersection of $y = 0.4x^2 - x + 5$ and $y = -3x^2 + 3x + 8$.

21. Use the distance formula to compute the distance between (2, 4) and (1, –2) to four decimals.

22. Use the midpoint formula to find the midpoint of the line segment joining (1, 3) and (–5, 6).

23. Find the center and radius of the circle having the equation $(x+6)^2 + (y-7)^2 = 25$.

24. Determine the equation of the circle having center (–7, 5) and radius 4.

25. Find the center and radius of the circle having the equation $x^2 + 2x + y^2 + 3y = 15$.

Test form 1E Date _________ Name ________________________

Course ______ Sec ______ Student Number ________________

Chapter 1

1. Determine which of the given points lie on the graph of $y = -4x^2 + 2x + 1$. (1, 1), (0, 1), (1, −1)

2. Find the x- and y-intercepts of the graph of $y = -3x - 18$ by letting y and then x be zero.

3. The points at which the line crosses the x and y axes are the integers 5 and −10, respectively. Find the slope of the line.

4. Use the point-slope form, $y - y_1 = m(x - x_1)$, with $m = \dfrac{y_2 - y_1}{x_2 - x_1}$ to find the equation of the line through the points (7, 8) and (4, 3).

5. Find the equation of the horizontal line through (6, 4).

6. Find the equation of the vertical line through (6, 4).

7. Find an equation of the line M through the point (3, 9) that is parallel to $15x - 3y = 9$. Give answer in point-slope form.

8. Find an equation of the line M through the point (2, 6) that is perpendicular to $15x - 3y = 9$. Give answer in point-slope form.

9. Use a calculator to determine the least squares line through (1, 8), (2, 13), (3, 20), and (4, 24).

10. Use a calculator to find the value of the linear function $f(x) = 6.84x - 9.21$ at $x = -7.32$

11. Complete the table for a linear function.

x	y
0	6
1	12
2	18
3	24
4	—

12. A refrigerator costs \$981 and there is a cost of \$60 per year to run it. Construct a function that describes the cost of owning and running the refrigerator.

13. Find the vertex of $y = 4x^2 + 9x - 1$.

14. Find the axis of symmetry of the graph $y = 4x^2 + 9x - 1$.

15. Find the *y*-intercept of the graph of $y = 4x^2 + 9x - 1$.

16. Find the *x*-intercepts of the parabola by factoring the equation $y = x^2 - 10x + 21$.

17. Find the *x*-intercepts of the graph of $y = 2x^2 - 6x - 17$.

18. Find *p* so that (1, 8) lies on the graph of $y = x^2 + 4px - 21$.

19. If an object is propelled vertically from the ground level with a velocity *v*, its motion in meters is described by the quadratic function $s(t) = -16t^2 + vt$. A particular object which is propelled reaches its maximum height in 12 seconds. What is the object's initial velocity?

20. Find the points of intersection of $y = 0.7x^2 - 2x + 5$ and $y = -3x^2 + 4x + 11$.

21. Use the distance formula to compute the distance between (2, 5) and (1, −11) to four decimals.

22. Use the midpoint formula to find the midpoint of the line segment joining (3, 13) and (−5, 8).

23. Find the center and radius of the circle having the equation $(x+9)^2 + (y-3)^2 = 49$.

24. Determine the equation of the circle having center (−8, 2) and radius 9.

25. Find the center and radius of the circle having the equation $x^2 + 6x + y^2 + 4y = 42$.

Test form 1F Date ________ Name ________________________
Course ______ Sec ______ Student Number ________________

Chapter 1

1. Determine which of the given points lie on the graph of $y = 2x^2 + 3x + 8$. (1, 8), (0, 8), (1, 13)

2. Find the x- and y-intercepts of the graph of $y = -3x + 12$ by letting y and then x be zero.

3. The points at which the line crosses the x and y axes are the integers 2 and -10, respectively. Find the slope of the line.

4. Use the point-slope form, $y - y_1 = m(x - x_1)$, with $m = \dfrac{y_2 - y_1}{x_2 - x_1}$ to find the equation of the line through the points (1, 8) and (5, 2).

5. Find the equation of the horizontal line through (4, 7).

6. Find the equation of the vertical line through (4, 7).

7. Find an equation of the line M through the point (4, 9) that is parallel to $21x + 3y = 6$. Give answer in point-slope form.

8. Find an equation of the line M through the point (2, 6) that is perpendicular to $21x + 3y = 6$. Give answer in point-slope form.

9. Use a calculator to determine the least squares line through (1, 5), (2, 7), (3, 10), and (4, 11).

10. Use a calculator to find the value of the linear function $f(x) = 4.14x - 8.24$ at $x = -3.76$

11. Complete the table for a linear function.

x	y
0	4
1	8
2	12
3	16
4	__

12. A refrigerator costs \$972 and there is a cost of \$61 per year to run it. Construct a function that describes the cost of owning and running the refrigerator.

13. Find the vertex of $y = 2x^2 + 5x - 6$.

14. Find the axis of symmetry of the graph $y = 2x^2 + 5x - 6$.

15. Find the *y*-intercept of the graph of $y = 2x^2 + 5x - 6$.

16. Find the *x*-intercepts of the parabola by factoring the equation $y = x^2 - 10x + 16$.

17. Find the *x*-intercepts of the graph of $y = 2x^2 - 3x - 10$.

18. Find *p* so that (1, 2) lies on the graph of $y = x^2 + 4px - 11$.

19. If an object is propelled vertically from the ground level with a velocity *v*, its motion in meters is described by the quadratic function $s(t) = -16t^2 + vt$. A particular object which is propelled reaches its maximum height in 9 seconds. What is the object's initial velocity?

20. Find the points of intersection of $y = 0.5x^2 - x + 5$ and $y = -3x^2 + 3x + 10$.

21. Use the distance formula to compute the distance between (9, 3) and (1, –1) to four decimals.

22. Use the midpoint formula to find the midpoint of the line segment joining (2, 7) and (–5, 3).

23. Find the center and radius of the circle having the equation $(x+3)^2 + (y-8)^2 = 36$.

24. Determine the equation of the circle having center (–9, 2) and radius 8.

25. Find the center and radius of the circle having the equation $x^2 + 3x + y^2 + 4y = 30$.

Test form 2A Date _________ Name _________________________
Course ______ Sec ______ Student Number ________________

Chapter 2

2-1. Solve the quadratic equation by factoring. $x^2 + 11x + 18 = 0$

A. $x = -18, 18$ B. $x = -3\sqrt{2}, 3\sqrt{2}$ C. $x = -2, -9$

D. $x = 2, 9$ E. $x = 3\sqrt{2}$

2-2. Solve the quadratic equation using the method of factoring. $3x^2 + 7x + 2 = 0$

A. $x = -\frac{1}{3}, -2$ B. $x = \frac{1}{3}, 2$ C. $x = -\frac{3}{2}, -1$ D. $x = \frac{3}{2}, 1$

E. $x = -\frac{1}{2}, 2$

2-3. Solve the quadratic equation to four decimal places, by extracting the square root. $x^2 = 7$

A. $x = -2.6458, 2.6458$ B. $x = -7, 7$ C. $x = 7$

D. $x = -2.6458$ E. $x = 2.6458$

2-4. Use the quadratic formula to solve the equation. Give answer to four decimal places. $2x^2 - 7x - 11 = 0$

A. $-1.1762, 4.6762$ B. 4.6762 C. -137

D. 137 E. $-137, 137$

2-5. Use the discriminant to determine how many real solutions the quadratic equation has. $x^2 - 2x + 5 = 0$

A. one real solution

B. two real solutions

C. no real solutions

D. The number of real solutions cannot be determined.

E. three real solutions

2-6. Use the equation $s = -16t^2 + vt$ that gives the height s feet at time t seconds of an object fired upward with a velocity v feet per second. If the object is propelled at 48 feet per second, when will it come back to the ground?

A. 6 seconds B. 1.5 seconds C. never

D. 3 seconds E. 9 seconds

2-7. A person wants to pave a 108-square foot patio. If the length is to be 3 feet greater than the width, find the dimensions of the patio.

A. 13 feet by 16 feet B. 12 feet by 15 feet C. 11 feet by 14 feet

D. 10 feet by 13 feet E. 9 feet by 12 feet

2-8. Solve $x^4 - 13x^2 + 36 = 0$ by using an appropriate substitution.

A. $2, 3$ B. $0, 2, 3$ C. $-2, 2, -3, 3$

E. $-9, 9, -4, 4$ F. $4, 9$

2-9. Solve the equation $x^{3/2} = 125$.

A. 5 B. $\frac{1}{25}$ C. $\frac{1}{125}$ D. $\frac{1}{5}$ E. 25

2-10. Solve the equation $x^{-2}+17x^{-1}+70=0$.

A. $-\frac{1}{70}$ B. $-\frac{1}{70}, \frac{1}{70}$ C. $-\frac{1}{7}, -\frac{1}{10}$

D. $\frac{1}{7}, \frac{1}{70}$ E. $\frac{1}{7}$

2-11. Solve the equation $x=\sqrt{-21x-38}$.

A. $-38, 38$ B. $-21, 21$ C. $-19, -2$

D. $2, 19$ E. No solution

2-12. Solve the equation $\sqrt{2x-9}=\sqrt{x+3}$.

A. 0 B. 12 C. -12

D. 90 E. No solution

2-13. Solve $x^3+2x^2-3x=0$.

A. $-3, 0, 1$ B. $-1, 0, 3$ C. $-3, 1$

D. $1, 3$ E. $-1, 3$

2-14. Solve the linear inequality. Write the solution in interval notation. $4x+5<3$

A. $\left(-\infty, -\frac{1}{2}\right)$ B. $\left(-\infty, -\frac{1}{2}\right]$ C. $\left(-\frac{1}{2}, \infty\right)$

D. $\left(-\frac{1}{2}, \frac{1}{2}\right)$ E. $\left[-\frac{1}{2}, \infty\right)$

2-15. Solve the linear inequality. Write the solution in interval notation.
$10x + 9 > 9x + 11$

A. $(-\infty, 2]$ B. $(2, \infty)$ C. $(-\infty, 2)$

D. $[2, \infty)$ E. $(-2, 2)$

2-16. Solve the linear inequality. Write the solution in interval notation. $8x + 4 > 9x + 6$

A. $(-\infty, -2]$ B. $(-2, 2)$ C. $(-2, \infty)$

D. $[-2, \infty)$ E. $(-\infty, -2)$

2-17. Solve the inequality $6 \le 2x + 4 \le 18$.

A. $(1, 7)$ B. $(-\infty, 1) \cup (7, \infty)$ C. $(-\infty, 1] \cup [7, \infty)$

D. $[1, 7]$ E. $(-\infty, 18)$

2-18. Solve the quadratic inequality $x^2 + 18x + 65 < 0$.

A. $(-\infty, -13)$ B. $(-13, -5)$

C. $(-\infty, -13) \cup (-5, \infty)$ D. $[-13, -5]$

E. $(-5, \infty)$

2-19. Solve the quadratic inequality $2x^2 + 38x \ge -168$.

A. $(-\infty, -12] \cup [-7, \infty)$ B. $(-12, -7)$

C. $(-\infty, -12)$ D. $(-7, \infty)$

E. $[-12, -7]$

2-20. Solve the inequality $|x| = 9$.

A. $-9, 9$ B. -9 C. 9

D. $-9, 0, 9$ E. $0, 9$

2-21. Solve the inequality $|x+11|=6$

A. 5, 17 B. −17, 6 C. −6

D. −5, −17 E. 6

2-22. Solve the inequality. Write the solution in interval notation. $|x|<8$

A. [−8, 8] B. (−8, 8) C. (−∞, 8)

D. (−∞, −8) ∪ (8, ∞) E. (0, 8)

2-23. Solve the inequality. Write the solution in interval notation. $|x|\geq 7$

A. (−∞, −7] ∪ [7, ∞) B. [−7, 7] C. (−∞, 7)

D. (−7, 7) E. (0, 7)

2-24. Solve the inequality. Write the solution in interval notation. $|x+2|<12$

A. (−12, 12) B. (−14, 10)

C. (−∞, −12) ∪ (12, ∞) D. (−∞, −14) ∪ (12, ∞)

E. (0, 10)

2-25. Solve the inequality. Write the solution in interval notation. $|x+2|>7$

A. (−7, 7) B. (−9, 5) C. (−∞, −9) ∪ (5, ∞)

D. (−∞, −7) ∪ (7, ∞) E. (0, 5)

Test form 2B Date __________ Name ____________________________
Course ______ Sec ______ Student Number __________________

Chapter 2

2-1. Solve the quadratic equation by factoring. $x^2 + 17x + 70 = 0$

A. $x = -70, 70$ B. $x = -\sqrt{70}, \sqrt{70}$ C. $x = 7, 10$

D. $x = -10, -7$ E. $x = \sqrt{70}$

2-2. Solve the quadratic equation using the method of factoring. $5x^2 + 21x + 4 = 0$

A. $x = \frac{1}{5}, 4$ B. $x = -\frac{1}{5}, -4$ C. $x = -\frac{4}{5}, -1$ D. $x = \frac{4}{5}, 1$

E. $x = -\frac{1}{5}, 5$

2-3. Solve the quadratic equation to four decimal places, by extracting the square root. $x^2 = 11$

A. $x = 11$ B. $x = -3.3166, 3.3166$ C. $x = -3.3166$

D. $x = 3.3166$ E. $x = -11, 11$

2-4. Use the quadratic formula to solve the equation. Give answer to four decimal places. $4x^2 - 6x - 13 = 0$

A. $-244, 244$ B. -244 C. 244

D. $-1.2026, 2.7026$ E. 2.7026

2-5. Use the discriminant to determine how many real solutions the quadratic equation has. $x^2 - 2x + 8 = 0$

A. one real solution
B. two real solutions
C. no real solutions
D. The number of real solutions cannot be determined.
E. three real solutions

2-6. Use the equation $s = -16t^2 + vt$ that gives the height s feet at time t seconds of an object fired upward with a velocity v feet per second. If the object is propelled at 64 feet per second, when will it come back to the ground?

A. 8 seconds B. 2 seconds C. 4 seconds

D. never E. 12 seconds

2-7. A person wants to pave a 130-square foot patio. If the length is to be 3 feet greater than the width, find the dimensions of the patio.

A. 13 feet by 12 feet B. 12 feet by 15 feet C. 11 feet by 14 feet

D. 10 feet by 13 feet E. 9 feet by 12 feet

2-8. Solve $x^4 - 25x^2 + 144 = 0$ by using an appropriate substitution.

A. 3, 4 B. 0, 3, 4 C. $-9, 9, -16, 16$

E. $-3, 3, -4, 4$ F. 9, 16

2-9. Solve the equation $x^{3/2} = 64$.

A. 4 B. 16 C. 64 D. $\frac{1}{16}$ E. $\frac{1}{64}$

2-10. Solve the equation $x^{-2} + 19x^{-1} + 88 = 0.$

A. $-\frac{1}{11}, -\frac{1}{8}$ B. $-18, 18$ C. $-\frac{1}{88}, \frac{1}{88}$

D. $\frac{1}{4}$ E. $-\frac{1}{88}$

2-11. Solve the equation $x = \sqrt{-13x - 30}.$

A. $-30, 30$ B. $-13, 13$ C. $-3, 10$

D. $3, 10$ E. No solution

2-12. Solve the equation $\sqrt{2x - 8} = \sqrt{x + 4}.$

A. 0 B. 12 C. -12

D. 80 E. No solution

2-13. Solve $x^3 + 2x^2 - 8x = 0.$

A. 2, 4 B. $-4, 2, 4$ C. 4

D. $-4, 0, 2$ E. $-4, 2$

2-14. Solve the linear inequality. Write the solution in interval notation. $3x + 5 < 3$

A. $\left(-\infty, -\frac{2}{3}\right)$ B. $\left(-\infty, -\frac{2}{3}\right]$ C. $\left(-\frac{2}{3}, \infty\right)$

D. $\left(-\frac{2}{3}, \frac{2}{3}\right)$ E. $\left[-\frac{2}{3}, \infty\right)$

2-15. Solve the linear inequality. Write the solution in interval notation.
$9x + 8 > 8x + 10$

A. $(-\infty, 2]$ B. $[2, \infty)$ C. $[-\infty, 2)$

D. $(2, \infty)$ E. $(-2, 2)$

2-16. Solve the linear inequality. Write the solution in interval notation. $7x + 3 > 8x + 5$

A. $(-\infty, -2]$ B. $(-2, 2)$ C. $(-2, \infty)$

D. $[-2, \infty)$ E. $(-\infty, -2)$

2-17. Solve the inequality $5 \leq 2x + 4 \leq 15$.

A. $\left[\frac{1}{2}, \frac{11}{2}\right]$ B. $\left(-\infty, \frac{1}{2}\right) \cup \left(\frac{11}{2}, \infty\right)$

C. $\left(-\infty, \frac{1}{2}\right] \cup \left[\frac{11}{2}, \infty\right)$ D. $\left(\frac{1}{2}, \frac{11}{2}\right)$

E. $(-\infty, 15)$

2-18. Solve the quadratic inequality $x^2 + 12x + 27 < 0$.

A. $(-\infty, -9)$ B. $(-9, -3)$

C. $(-\infty, -9) \cup (-3, \infty)$ D. $[-9, -3]$

E. $(-3, \infty)$

2-19. Solve the quadratic inequality $2x^2 + 24x \geq -70$.

A. $(-\infty, -5] \cup [-7, \infty)$ B. $[-7, -5]$

C. $(-\infty, -7)$ D. $(-7, -5)$

E. $(-5, \infty)$

2-20. Solve the inequality $|x| = 15$.

A. $-15, 15$ B. -15 C. 15

D. $-15, 0, 15$ E. $0, 15$

2-21. Solve the inequality $|x+9| = 4$.

A. $5, 13$ B. $-4, 4$ C. $-13, -5$

D. 4 E. -5

2-22. Solve the inequality. Write the solution in interval notation. $|x| < 3$

A. $[-3, 3]$ B. $(-3, 3)$ C. $(-\infty, 3)$

D. $(-\infty, -3) \cup (3, \infty)$ E. $(0, 3)$

2-23. Solve the inequality. Write the solution in interval notation. $|x| \geq 4$

A. $(-4, 4)$ B. $[-4, 4]$ C. $(-\infty, 4)$

D. $(-\infty, -4] \cup [4, \infty)$ E. $(0, 4)$

2-24. Solve the inequality. Write the solution in interval notation. $|x+2| < 6$

A. $(-8, 4)$ B. $(-6, 6)$

C. $(-\infty, -6) \cup (6, \infty)$ D. $(-\infty, -8) \cup (4, \infty)$

E. $(0, 4)$

2-25. Solve the inequality. Write the solution in interval notation. $|x+2| > 6$

A. $(-6, 6)$ B. $(-8, 4)$ C. $(-\infty, -6) \cup (6, \infty)$

D. $(-\infty, -8) \cup (4, \infty)$ E. $(0, 4)$

Test form 2C Date _________ Name _________________________
Course ______ Sec ______ Student Number ________________

Chapter 2

2-1. Solve the quadratic equation by factoring. $x^2 + 19x + 88 = 0$

A. $x = -11, -8$ B. $x = 8, 11$ C. $x = -2\sqrt{22}, 2\sqrt{22}$

D. $x = -88, 88$ E. $x = 2\sqrt{22}$

2-2. Solve the quadratic equation using the method of factoring. $3x^2 + 22x + 7 = 0$

A. $x = -\frac{7}{3}, -1$ B. $x = \frac{7}{3}, 1$ C. $x = -\frac{1}{3}, -7$ D. $x = \frac{1}{3}, 7$

E. $x = -\frac{1}{3}, 7$

2-3. Solve the quadratic equation to four decimal places, by extracting the square root. $x^2 = 15$

A. $x = 3.8730$ B. $x = -3.8730, 3.8730$ C. $x = 15$

D. $x = -15, 15$ E. $x = -3.8730$

2-4. Use the quadratic formula to solve the equation. Give answer to four decimal places. $3x^2 - 7x - 13 = 0$

A. 3.5530 B. -1.2196, 3.5530 C. 205

D. -205 E. -205, 205

2-5. Use the discriminant to determine how many real solutions the quadratic equation has. $x^2 - 2x + 54 = 0$

A. one real solution

B. two real solutions

C. no real solutions

D. The number of real solutions cannot be determined.

E. three real solutions

2-6. Use the equation $s = -16t^2 + vt$ that gives the height s feet at time t seconds of an object fired upward with a velocity v feet per second. If the object is propelled at 80 feet per second, when will it come back to the ground?

A. 5 seconds B. 2.5 seconds C. 25 seconds

D. never E. 10 seconds

2-7. A person wants to pave a 154-square foot patio. If the length is to be 3 feet greater than the width, find the dimensions of the patio.

A. 13 feet by 12 feet B. 12 feet by 15 feet C. 11 feet by 14 feet

D. 10 feet by 13 feet E. 9 feet by 12 feet

2-8. Solve $x^4 - 41x^2 + 400 = 0$ by using an appropriate substitution.

A. $4, 5$ B. $-4, 4, -5, 5$ C. $0, 4, 5$

E. $-16, 16, -25, 25$ F. $16, 25$

2-9. Solve the equation $x^{3/2} = 216$.

A. 36 B. 6 C. 196 D. $\frac{1}{6}$ E. $\frac{1}{36}$

2-10. Solve the equation $x^{-2}+11x^{-1}+18=0$.

A. $\frac{1}{2}$ B. $-\frac{1}{2}, -\frac{1}{9}$ C. $-\frac{1}{18}, \frac{1}{18}$

D. $\frac{1}{2}, \frac{1}{9}$ E. $-\frac{1}{18}$

2-11. Solve the equation $x=\sqrt{-14x-40}$.

A. $-40, 40$ B. $-14, 14$ C. $-4, -10$

D. $4, 10$ E. No solution

2-12. Solve the equation $\sqrt{2x-6}=\sqrt{x+4}$.

A. 0 B. -10 C. 10

D. 24 E. No solution

2-13. Solve $x^3+x^2-12x=0$.

A. 3, 5 B. $-4, 0, 3$ C. $-3, 0$

D. $-3, 4$ E. 0, 3, 4

2-14. Solve the linear inequality. Write the solution in interval notation. $7x+5<3$

A. $\left(-\frac{2}{7}, \infty\right)$ B. $\left(-\infty, -\frac{2}{7}\right]$ C. $\left(-\infty, -\frac{2}{7}\right)$

D. $\left(-\frac{2}{7}, \frac{2}{7}\right)$ E. $\left[-\frac{2}{7}, \infty\right)$

2-15. Solve the linear inequality. Write the solution in interval notation.
$7x+5>6x+7$

A. $(-\infty, 2]$ B. $[-\infty, 2)$ C. $(-2, 2)$

D. $[2, \infty)$ E. $(2, \infty)$

2-16. Solve the linear inequality. Write the solution in interval notation. $5x+3<6x+5$

A. $(-\infty, -2]$ B. $(-2, 2)$ C. $(-2, \infty)$

D. $(-\infty, -2)$ E. $[-2, \infty)$

2-17. Solve the inequality $7 \leq 2x+4 \leq 25$.

A. $\left[\frac{3}{2}, \frac{21}{2}\right]$ B. $\left(-\infty, \frac{3}{2}\right) \cup \left(\frac{21}{2}, \infty\right)$

C. $\left(-\infty, \frac{3}{2}\right] \cup \left[\frac{21}{2}, \infty\right)$ D. $\left(\frac{3}{2}, \frac{21}{2}\right)$

E. $(-\infty, 21)$

2-18. Solve the quadratic inequality $x^2+9x+20<0$.

A. $(-\infty, -5) \cup (-4, \infty)$ B. $(-5, -4)$

C. $(-\infty, -4)$ D. $[-5, -4]$

E. $(-5, \infty)$

2-19. Solve the quadratic inequality $2x^2+30x \geq -100$.

A. $(-\infty, -10] \cup [-5, \infty)$ B. $(-10, -5)$

C. $(-\infty, -1)$ D. $[-10, -5]$

E. $(-5, \infty)$

2-20. Solve the inequality $|x| = 17$.

A. $-17, 17$ B. -17 C. 17

D. $-17, 0, 17$ E. $0, 17$

2-21. Solve the inequality $|x+8| = 13$.

A. $-5, 21$ B. $-21, 5$ C. $-8, 8$

D. 5 E. 8

2-22. Solve the inequality. Write the solution in interval notation. $|x| < 12$

A. $(-12, 12)$ B. $[-12, 12]$ C. $(-\infty, 12)$

D. $(-\infty, -12) \cup (12, \infty)$ E. $(0, 12)$

2-23. Solve the inequality. Write the solution in interval notation. $|x| \geq 16$

A. $(-16, 16)$ B. $[-16, 16]$ C. $(-\infty, 16)$

D. $(-\infty, -16] \cup [16, \infty)$ E. $(0, 16)$

2-24. Solve the inequality. Write the solution in interval notation. $|x+2| < 7$

A. $(-\infty, -9) \cup (5, \infty)$ B. $(-7, 7)$

C. $(-\infty, -7) \cup (7, \infty)$ D. $(-9, 5)$

E. $(0, 5)$

2-25. Solve the inequality. Write the solution in interval notation. $|x+2| > 4$

A. $(-4, 4)$ B. $(-6, 2)$ C. $(-\infty, -4) \cup (4, \infty)$

D. $(0, 2)$ E. $(-\infty, -6) \cup (2, \infty)$

Test form 2D Date _________ Name _________________________
Course ______ Sec ______ Student Number ________________

Chapter 2

2-1. Solve the quadratic equation by factoring. $x^2 + 23x + 42 = 0$

2-2. Solve the quadratic equation using the method of factoring. $5x^2 + 41x + 8 = 0$

2-3. Solve the quadratic equation to four decimal places, by extracting the square root. $x^2 = 6$

2-4. Use the quadratic formula to solve the equation. Give answer to four decimal places. $2x^2 - 7x - 13 = 0$

2-5. Use the discriminant to determine how many real solutions the quadratic equation has. $x^2 - 2x + 9 = 0$

2-6. Use the equation $s = -16t^2 + vt$ that gives the height s feet at time t seconds of an object fired upward with a velocity v feet per second. If the object is propelled at 96 feet per second, when will it come back to the ground?

2-7. A person wants to pave a 208-square foot patio. If the length is to be 3 feet greater than the width, find the dimensions of the patio.

2-8. Solve $x^4 - 5x^2 + 4 = 0$ by using an appropriate substitution.

2-9. Solve the equation $x^{3/4} = 125$.

2-10. Solve the equation $x^{-2} + 13x^{-1} + 30 = 0$.

2-11. Solve the equation $x = \sqrt{-15x - 50}$.

2-12. Solve the equation $\sqrt{2x - 5} = \sqrt{x + 7}$.

2-13. Solve $x^3 - 2x^2 - 8x = 0$.

2-14. Solve the linear inequality. Write the solution in interval notation. $5x + 7 < 9$

2-15. Solve the linear inequality. Write the solution in interval notation.
$5x+10>4x+12$

2-16. Solve the linear inequality. Write the solution in interval notation. $4x+2<5x+4$

2-17. Solve the inequality $6\leq 2x+4\leq 20$.

2-18. Solve the quadratic inequality $x^2+10x+21<0$.

2-19. Solve the quadratic inequality $2x^2+46x\leq -152$.

2-20. Solve the inequality $|x|=27$.

2-21. Solve the inequality $|x+4|=20$.

2-22. Solve the inequality. Write the solution in interval notation. $|x|<13$

2-23. Solve the inequality. Write the solution in interval notation. $|x|\geq 23$

2-24. Solve the inequality. Write the solution in interval notation. $|x+2|<6$

2-25. Solve the inequality. Write the solution in interval notation. $|x+2|>6$

Test form 2E Date ________ Name ______________________

Course ______ Sec ______ Student Number ________________

Chapter 2

2-1. Solve the quadratic equation by factoring. $x^2 + 13x + 30 = 0$

2-2. Solve the quadratic equation using the method of factoring. $5x^2 + 46x + 9 = 0$

2-3. Solve the quadratic equation to four decimal places, by extracting the square root. $x^2 = 14$

2-4. Use the quadratic formula to solve the equation. Give answer to four decimal places. $2x^2 - 8x - 13 = 0$

2-5. Use the discriminant to determine how many real solutions the quadratic equation has. $x^2 - 4x + 4 = 0$

2-6. Use the equation $s = -16t^2 + vt$ that gives the height s feet at time t seconds of an object fired upward with a velocity v feet per second. If the object is propelled at 112 feet per second, when will it come back to the ground?

2-7. A person wants to pave a 460-square foot patio. If the length is to be 3 feet greater than the width, find the dimensions of the patio.

2-8. Solve $x^4 - 10x^2 + 9 = 0$ by using an appropriate substitution.

2-9. Solve the equation $x^{3/4} = 8$.

2-10. Solve the equation $x^{-2} + 15x^{-1} + 50 = 0$.

2-11. Solve the equation $x = \sqrt{-16x - 60}$.

2-12. Solve the equation $\sqrt{2x - 8} = \sqrt{x + 9}$.

2-13. Solve $x^3 - 8x^2 + 7x = 0$.

2-14. Solve the linear inequality. Write the solution in interval notation. $2x + 5 < 3$

2-15. Solve the linear inequality. Write the solution in interval notation.
$7x+12>6x+14$

2-16. Solve the linear inequality. Write the solution in interval notation. $5x+3<6x+5$

2-17. Solve the inequality $2\le 2x+4\le 12$.

2-18. Solve the quadratic inequality $x^2+13x+22<0$.

2-19. Solve the quadratic inequality $2x^2+30x\ge 200$.

2-20. Solve the inequality $|x|=29$.

2-21. Solve the inequality $|x+2|=30$.

2-22. Solve the inequality. Write the solution in interval notation. $|x|<60$

2-23. Solve the inequality. Write the solution in interval notation. $|x|\ge 10$

2-24. Solve the inequality. Write the solution in interval notation. $|x+2|<50$

2-25. Solve the inequality. Write the solution in interval notation. $|x+2|>3$

Test form 2F Date __________ Name __________________________
Course ______ Sec ______ Student Number ________________

Chapter 2

2-1. Solve the quadratic equation by factoring. $x^2 + 15x + 50 = 0$

2-2. Solve the quadratic equation using the method of factoring. $3x^2 + 25x + 8 = 0$

2-3. Solve the quadratic equation to four decimal places, by extracting the square root. $x^2 = 3$

2-4. Use the quadratic formula to solve the equation. Give answer to four decimal places. $4x^2 - 7x - 17 = 0$

2-5. Use the discriminant to determine how many real solutions the quadratic equation has. $x^2 - x - 8 = 0$

2-6. Use the equation $s = -16t^2 + vt$ that gives the height s feet at time t seconds of an object fired upward with a velocity v feet per second. If the object is propelled at 128 feet per second, when will it come back to the ground?

2-7. A person wants to pave a 550-square foot patio. If the length is to be 3 feet greater than the width, find the dimensions of the patio.

2-8. Solve $x^4 - 17x^2 + 16 = 0$ by using an appropriate substitution.

2-9. Solve the equation $x^{3/4} = 27$

2-10. Solve the equation $x^{-2} + 18x^{-1} + 80 = 0$.

2-11. Solve the equation $x = \sqrt{-15x - 54}$.

2-12. Solve the equation $\sqrt{2x - 1} = \sqrt{x + 4}$.

2-13. Solve $x^3 + 2x^2 - 35x = 0$.

2-14. Solve the linear inequality. Write the solution in interval notation. $5x + 10 < 15$

2-15. Solve the linear inequality. Write the solution in interval notation.
$3x + 15 > 2x + 17$

2-16. Solve the linear inequality. Write the solution in interval notation.
$3x + 17 < 4x + 19$

2-17. Solve the inequality $5 \leq 2x + 4 \leq 15$.

2-18. Solve the quadratic inequality $x^2 + 19x + 48 < 0$.

2-19. Solve the quadratic inequality $2x^2 + 38x \geq -120$.

2-20. Solve the inequality $|x| = 32$.

2-21. Solve the inequality $|x + 1| = 9$.

2-22. Solve the inequality. Write the solution in interval notation. $|x| < 47$

2-23. Solve the inequality. Write the solution in interval notation. $|x| \geq 52$

2-24. Solve the inequality. Write the solution in interval notation. $|x + 2| < 19$

2-25. Solve the inequality. Write the solution in interval notation. $|x + 2| > 21$

Test form 3A   Date _________   Name ________________________
Course ______ Sec ______   Student Number ________________

Chapter 3

3-1. Determine the x-intercepts of the graph of the polynomial function $f(x) = x(x-3)(x+8)$

A. $-3, 0, 8$ B. $0, 3, 8$ C. $-8, 0, 3$ D. $-3, 8$ E. $-8, 3$

3-2. Find the polynomial function that has x-intercepts –4, 4, and 5, and y-intercept 160.

A. $f(x) = x^3 - 5x^2 - 16x - 10$ B. $f(x) = 2x^3 + 10x^2 + 40x + 160$

C. $f(x) = 2x^3 - 10x^2 - 32x + 160$ D. $f(x) = x^3 + 10x^2 + 20x + 80$

E. $f(x) = x^3 - 4x^2 - 4x - 5$

3-3. Determine where the quadratic function $f(x) = -x^2 + 10x - 30$ is increasing.

A. $(-\infty, 5)$ B. $(-\infty, -5)$ C. $(-\infty, 10)$ D. $(5, \infty)$ E. $(-5, \infty)$

3-4. Find the local minimum for $f(x) = x^2 + 8x + 12$.

A. 8 B. –2 C. 14 D. 60 E. –4

3-5. Determine the local maximum for $f(x) = 2x^3 + 4x^2 - 6x + 10$ to four decimal places.

A. 2 B. 12.1292 C. 22.1292 D. 12 E. 72.7753

3-6. Determine the domain and range of $f(x) = x^2 + 4x + 4$.

A. Domain: $(-\infty, \infty)$; Range: $[2, \infty)$ B. Domain: $(-\infty, \infty)$; Range: $[0, \infty)$

C. Domain: $(-\infty, \infty)$; Range: $(0, \infty)$ D. Domain: $[0, \infty)$; Range: $(-\infty, \infty)$

E. Domain: $(0, \infty)$; Range: $(-\infty, \infty)$

3-7. Which of these functions are even: $f(x) = 2x^3$, $g(x) = 4x^2$, $h(x) = 6|x^3|$?

A. $f(x)$ only

B. $g(x)$ only

C. $g(x)$ and $h(x)$ only

D. $h(x)$ only

E. all three

3-8. Determine the symmetry of $f(x) = 4x^3 + 2x^5$.

A. Symmetric about the origin

B. Symmetric about the x-axis

C. Symmetric about the y-axis

D. Symmetric about the line $y = x$

E. Symmetric about the line $y = -x$

3-9. The graph of $g(x) = -4(x+2)^2 - 3$ is the graph of $f(x) = x^2$ with what transformations?

A. Horizontal shift of 2 units to the right, reflection about the x-axis, stretching by a factor of 4, and vertical shift of 3 units down

B. Horizontal shift of 2 units to the left, reflection about the y-axis, stretching by a factor of 4, and vertical shift of 3 units down

C. Horizontal shift of 2 units to the left, reflection about the x-axis, stretching by a factor of 4, and vertical shift of 3 units down

D. Horizontal shift of 2 units to the left, reflection about the y-axis, stretching by a factor of 4, and vertical shift of 3 units up

E. Horizontal shift of 2 units to the right, reflection about the y-axis, stretching by a factor of 4, and vertical shift of 3 units down

3-10. Rewrite the quadratic function $g(x) = x^2 - 4x + 1$ in a form that shows the shifts.

A. $(x+2)^2 + 5$ B. $(x+2)^2 - 5$ C. $(x-2)^2 - 5$

D. $(x-2)^2 + 5$ E. $(x-2)^2 - 5$

3-11. The graph shown here is the graph of

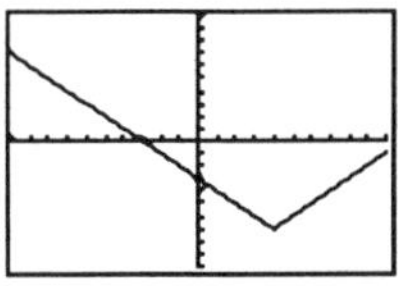

A. $f(x) = |x+4| - 7$ B. $f(x) = |x-4| - 7$ C. $f(x) = |x+4| + 7$

D. $f(x) = |x-4| - 7$ E. $f(x) = |x-4| + 7$

3-12 Determine the domain of $f(x) = \dfrac{(3x+4)(x-9)}{x-9}$.

A. All real numbers

B. All real numbers except –9

C. All real numbers except 9

D. All real numbers except $\dfrac{4}{3}$

E. All real numbers except $-\dfrac{4}{3}$

3-13. Give the equation for all vertical asymptotes of $f(x) = \dfrac{x-4}{8x+9}$.

A. $x = -\dfrac{9}{8}$ B. $x = 4$ C. $y = -\dfrac{1}{8}$ D. $y = \dfrac{1}{8}$ E. $x = \dfrac{9}{8}$

3-14. Give the equation for all horizontal asymptotes of $f(x)=\dfrac{x-7}{2x+9}$.

A. $x=-\dfrac{9}{2}$ B. $x=7$ C. $y=-\dfrac{1}{2}$ D. $y=\dfrac{1}{2}$ E. $x=\dfrac{9}{2}$

3-15. Give the equation for all vertical asymptotes of the function graphed here.

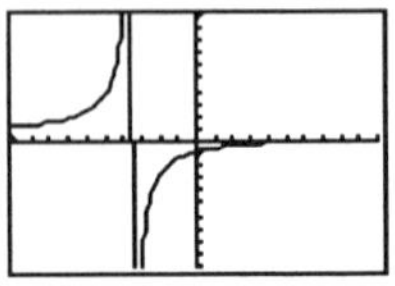

A. $x=-\dfrac{7}{2}$ B. $x=6$ C. $y=-\dfrac{1}{6}$ D. $y=\dfrac{1}{6}$ E. $x=\dfrac{7}{2}$

3-16. Give the equations of the asymptotes of $f(x)=x-9+\dfrac{7}{x+3}$.

A. Slant asymptote $y=x-9$

B. Slant asymptote $y=x-9$; horizontal asymptote $y=7$

C. Horizontal asymptote $y=9$

D. Slant asymptote $y=x-9$; vertical asymptote $x=-3$

E. Vertical asymptote $x=-3$

3-17. If $f(x)=\sqrt{x-4}$ and $g(x)=x+9$, write the expression for $(f+g)(x)$.

A. $\sqrt{2x+5)x}$

B. $\sqrt{x-4}+x+9$

C. $\sqrt{2x-5}$

D. $\sqrt{2x+5}$

E. $\sqrt{x}+x$

3-18. $f(x) = x^2 + 8$ and $g(x) = \frac{-4}{x}$. Find $(g \circ f)(x)$.

A. $\frac{16}{x^2} + 8$ B. $\frac{4}{x^2 + 8}$ C. $\frac{-4}{x^2 + 8}$

D. $\frac{-4(x^2 + 8)}{x}$ E. $x^2 + \frac{-32}{x}$

3-19. A company has a total cost function of a commodity over a year given by $C(x) = \frac{x^3}{100} + 15x^2 + 5000$ and a revenue function given by $R(x) = +18x^2 + 5000x$. If the profit function is $P(x) = R(x) - C(x)$, find $P(x)$.

A. $-2x^2 + 5000x$ B. $\frac{x^3}{100} - 7x^2 + 5000x + 5000$

C. $-\frac{x^3}{100} - 7x^2 + 5000x - 5000$ D. $\frac{x^3}{100} + 3x^2 + 5000x + 5000$

E. $-\frac{x^3}{100} + 3x^2 + 5000x - 5000$

3-20. $f(x) = x^3 + 1$; $g(x) = \sqrt{x + 2}$. Find $(f \circ g)(23)$.

A. 126 B. 23 C. 2 D. 60,840 E. 110.32

3-21. Of $f(x) = 4x^2$, $g(x) = 7x^3$, and $h(x) = \sqrt{x - 2}$, list all one-to-one functions.

A. $f(x)$ B. $g(x)$ C. $h(x)$

D. $g(x)$ and $h(x)$ E. $f(x), g(x)$, and $h(x)$

3-22. Of $f(x), g(x)$ and $h(x)$, list all one-to-one functions.

$f(x)$ 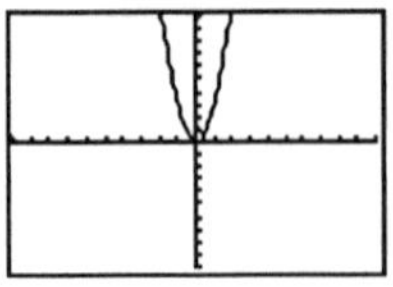

$g(x)$ 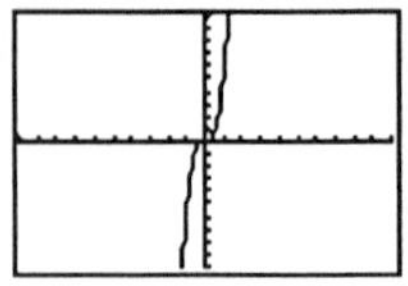

$h(x)$ 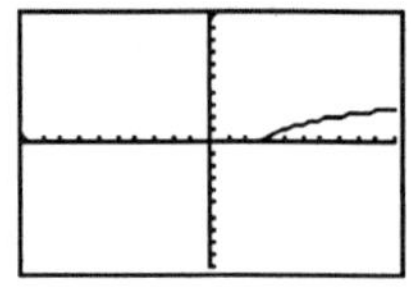

A. $f(x)$

B. $g(x)$

C. $h(x)$

D. $g(x)$ and $h(x)$

E. $f(x), g(x)$, and $h(x)$

3-23. Find the inverse function for $f(x) = 3x - 8$.

A. $f^{-1}(x) = \dfrac{1}{3x} - 8$

B. $f^{-1}(x) = \dfrac{1}{3x} + 8$

C. $f^{-1}(x) = \dfrac{1}{3x + 8}$

D. $f^{-1}(x) = \dfrac{x+8}{3}$

E. $f^{-1}(x) = \dfrac{x}{3} + 8$

3-24. Find the inverse function for $f(x) = x^5 + 3$.

A. $f^{-1}(x) = \dfrac{1}{x^5 + 3}$

B. $f^{-1}(x) = \dfrac{1}{x^5} + 3$

C. $f^{-1}(x) = x^5 + \dfrac{1}{3}$

D. $f^{-1}(x) = \sqrt[5]{x - 3}$

E. $f^{-1}(x) = \sqrt[5]{x + 3}$

3-25. If the velocity of a particle is $v(t) = \frac{t^3}{5}$, what is the expression for t as a function of v?

A. $\sqrt[3]{5t}$ B. $\sqrt[3]{5v}$ C. $\frac{5}{t^3}$ D. $5\sqrt[3]{v}$ E. $5\sqrt[3]{t}$

Test form 3B Date _________ Name _________________________
Course ______ Sec ______ Student Number ________________

Chapter 3

3-1. Determine the x-intercepts of the graph of the polynomial function $f(x) = x(x-2)(x+5)$

A. $-5, 0, 2$ B. $-2, 0, 5$ C. $0, 2, 5$ D. $-2, 5$ E. $-5, 2$

3-2. Find the polynomial function that has x-intercepts –4, 3, and 10, and y-intercept 240.

A. $f(x) = x^3 - 9x^2 + 22x - 120$ B. $f(x) = 2x^3 + 18x^2 + 44x + 240$

C. $f(x) = x^3 - 2x^2 - 3x - 10$ D. $f(x) = x^3 + 9x^2 + 22x + 120$

E. $f(x) = 2x^3 - 18x^2 - 44x + 240$

3-3. Determine where the quadratic function $f(x) = -x^2 + 10x - 16$ is increasing.

A. $(-\infty, 10)$ B. $(-\infty, -5)$ C. $(-\infty, 5)$ D. $(5, \infty)$ E. $(-5, \infty)$

3-4. Find the local minimum for $f(x) = x^2 + 10x + 21$.

A. 6 B. –4 C. 5 D. 10 E. –5

3-5. Determine the local maximum for $f(x) = 2x^3 + 4x^2 - 6x + 7$ to four decimal places.

A. 2 B. 12.1292 C. 19.1292 D. 9 E. 54.5814

3-6. Determine the domain and range of $f(x) = x^2 - 8x + 16$.

A. Domain: $(-\infty, \infty)$; Range: $[0, \infty)$ B. Domain: $(-\infty, \infty)$; Range: $[2, \infty)$

C. Domain: $(-\infty, \infty)$; Range: $(0, \infty)$ D. Domain: $[0, \infty)$; Range: $(-\infty, \infty)$

E. Domain: $(0, \infty)$; Range: $(-\infty, \infty)$

3-7. Which of these functions are even: $f(x)=8x^3$, $g(x)=3x^2$, $h(x)=7|x^3|$?

A. $f(x)$ only

B. $g(x)$ only

C. $g(x)$ and $h(x)$ only

D. $h(x)$ only

E. all three

3-8. Determine the symmetry of $f(x)=4x^3+7x^5$.

A. Symmetric about the x-axis

B. Symmetric about the origin

C. Symmetric about the y-axis

D. Symmetric about the line $y=x$

E. Symmetric about the line $y=-x$

3-9. The graph of $g(x)=-5(x+4)^2-6$ is the graph of $f(x)=x^2$ with what transformations?

A. Horizontal shift of 4 units to the right, reflection about the x-axis, stretching by a factor of 5, and vertical shift of 6 units down

B. Horizontal shift of 4 units to the left, reflection about the x-axis, stretching by a factor of 5, and vertical shift of 6 units down

C. Horizontal shift of 4 units to the left, reflection about the y-axis, stretching by a factor of 5, and vertical shift of 6 units up

D. Horizontal shift of 4 units to the right, reflection about the y-axis, stretching by a factor of 5, and vertical shift of 6 units down

E. Horizontal shift of 4 units to the left, reflection about the y-axis, stretching by a factor of 5, and vertical shift of 6 units down

3-10. Rewrite the quadratic function $g(x) = x^2 - 4x - 5$ in a form that shows the shifts.

A. $(x+2)^2 + 29$ B. $(x+4)^2 - 29$ C. $(x-2)^2 - 5$

D. $(x-2)^2 - 9$ E. $(x-2)^2 + 9$

3-11. The graph shown here is the graph of

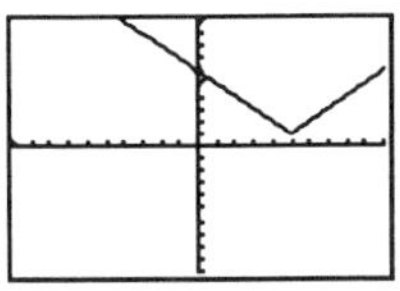

A. $f(x) = |x-5| + 1$ B. $f(x) = |x-5| - 1$ C. $f(x) = |x+5| + 1$

D. $f(x) = |x-5| - 1$ E. $f(x) = |x-1| + 5$

3-12. Determine the domain of $f(x) = \dfrac{(2x+5)(x-1)}{x-1}$.

A. All real numbers except 1 B. All real numbers except -1

C. All real numbers D. All real numbers except $\frac{5}{2}$

E. All real numbers except $-\frac{5}{2}$

3-13. Give the equation for all vertical asymptotes of $f(x) = \dfrac{x-2}{4x+7}$.

A. $x = -\frac{7}{4}$ B. $x = 2$ C. $y = -\frac{1}{4}$ D. $y = \frac{1}{4}$ E. $x = \frac{7}{4}$

3-14. Give the equation for all horizontal asymptotes of $f(x) = \dfrac{x-3}{2x+9}$.

A. $y = \frac{1}{2}$ B. $x = 3$ C. $y = -\frac{1}{2}$ D. $x = -\frac{2}{9}$ E. $x = \frac{2}{9}$

3-15. Give the equation for all vertical asymptotes of the function graphed here.

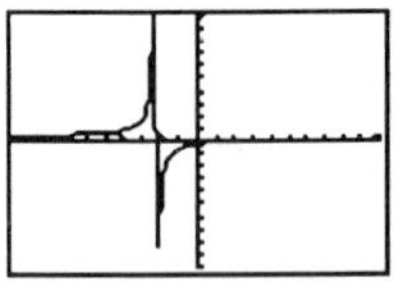

A. $y=-\frac{1}{4}$ B. $x=2$ C. $x=-\frac{9}{4}$ D. $y=\frac{1}{4}$ E. $x=\frac{9}{4}$

3-16. Give the equations of the asymptotes of $f(x)=x-2+\frac{3}{x+5}$.

A. Slant asymptote $y=x-2$

B. Slant asymptote $y=x-2$; horizontal asymptote $y=3$

C. Horizontal asymptote $y=2$

D. Slant asymptote $y=x-2$; vertical asymptote $x=-5$

E. Vertical asymptote $x=-5$

3-17. If $f(x)=\sqrt{x-3}$ and $g(x)=x+7$, write the expression for $(f+g)(x)$.

A. $\sqrt{(2x+4)x}$

B. $\sqrt{2x-4}$

C. $\sqrt{x-3}+x+7$

D. $\sqrt{2x+4}$

E. $\sqrt{x}+x$

3-18. $f(x)=x^2+1$ and $g(x)=\frac{2}{x}$. Find $(g \circ f)(x)$.

A. $\frac{4}{x^2}+1$ B. $\frac{2}{x^2+1}$ C. $\frac{-2}{x^2+1}$

D. $\frac{2(x^2+1)}{x}$ E. $x^2+\frac{-2}{x}$

3-19. A company has a total cost function of a commodity over a year given by $C(x)=\frac{x^3}{100}+14x^2+5000$ and a revenue function given by $R(x)=+17x^2+5000x$. If the profit function is $P(x)=R(x)-C(x)$, find $P(x)$.

A. $-2x^2+5000x$

B. $\frac{x^3}{100}-7x^2+5000x+5000$

C. $-\frac{x^3}{100}-7x^2+5000x-5000$

D. $-\frac{x^3}{100}+3x^2+5000x-5000$

E. $\frac{x^3}{100}+3x^2+5000x+5000$

3-20. $f(x)=x^3+1$; $g(x)=\sqrt{x+2}$. Find $(f\circ g)(34)$.

A. 216
B. 14.80
C. 318.75
D. 217
E. 34

3-21. Of $f(x)=x^2$, $g(x)=10x^3$, and $h(x)=\sqrt{x-2}$, list all one-to-one functions.

A. $f(x)$

B. $g(x)$

C. $h(x)$

D. $g(x)$ and $h(x)$

E. $f(x), g(x)$, and $h(x)$

3-22. Of $f(x), g(x)$ and $h(x)$, list all one-to-one functions.

$f(x)$

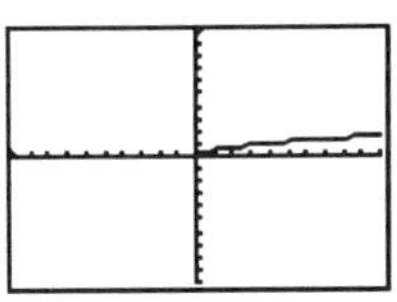

$g(x)$

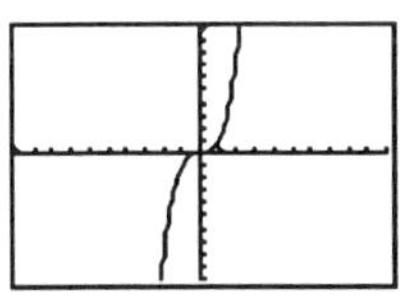

$h(x)$

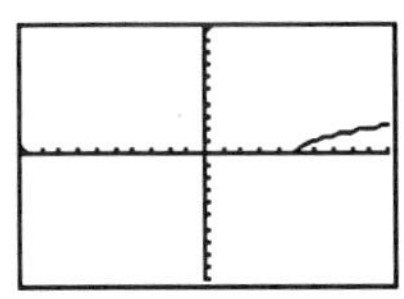

A. $f(x)$

B. $g(x)$

C. $h(x)$

D. $g(x)$ and $h(x)$

E. $f(x), g(x)$, and $h(x)$

3-23. Find the inverse function for $f(x) = 6x - 3$.

A. $f^{-1}(x) = \dfrac{1}{6x-3}$

B. $f^{-1}(x) = \dfrac{1}{6x} - 3$

C. $f^{-1}(x) = \dfrac{x+3}{6}$

D. $f^{-1}(x) = \dfrac{1}{6x} + 3$

E. $f^{-1}(x) = \dfrac{x}{6} + 3$

3-24. Find the inverse function for $f(x) = x^3 + 7$.

A. $f^{-1}(x) = \dfrac{1}{x^3 + 7}$

B. $f^{-1}(x) = \dfrac{1}{x^3} + 7$

C. $f^{-1}(x) = x^3 + \dfrac{1}{7}$

D. $f^{-1}(x) = \sqrt[3]{x+7}$

E. $f^{-1}(x) = \sqrt[3]{x-7}$

3-25. If the velocity of a particle is $v(t) = \dfrac{t^3}{20}$, what is the expression for t as a function of v?

A. $20\sqrt[3]{t}$ B. $\sqrt[3]{20t}$ C. $\dfrac{20}{t^3}$ D. $20\sqrt[3]{v}$ E. $\sqrt[3]{20v}$

Test form 3C Date ________ Name ________________________
Course ______ Sec ______ Student Number ________________

Chapter 3

3-1. Determine the x-intercepts of the graph of the polynomial function $f(x) = x(x-3)(x+4)$

A. $-3, 0, 4$ B. $-4, 3$ C. $-3, 4$ D. $-4, 0, 3$ E. $0, 3, 4$

3-2. Find the polynomial function that has x-intercepts -5, 3, and 6, and y-intercept 180.

A. $f(x) = x^3 - 4x^2 + 27x - 90$ B. $f(x) = 2x^3 + 8x^2 + 54x + 180$

C. $f(x) = x^3 + 4x^2 + 27x + 90$ D. $f(x) = 2x^3 - 8x^2 - 54x + 180$

E. $f(x) = x^3 - 5x^2 - 3x - 6$

3-3. Determine where the quadratic function $f(x) = -x^2 + 12x - 32$ is increasing.

A. $(-\infty, 12)$ B. $(-\infty, -6)$ C. $(-\infty, 6)$ D. $(6, \infty)$ E. $(-6, \infty)$

3-4. Find the local minimum for $f(x) = x^2 + 12x + 20$.

A. 12 B. -6 C. -16 D. 128 E. 6

3-5. Determine the local maximum for $f(x) = 2x^3 + 4x^2 - 6x + 1$ to four decimal places.

A. 13.1292 B. 12.1292 C. 2 D. 3 E. 18.1938

3-6. Determine the domain and range of $f(x) = x^2 - 4x + 4$.

A. Domain: $(-\infty, \infty)$; Range: $[2, \infty)$ B. Domain: $[0, \infty)$; Range: $(-\infty, \infty)$

C. Domain: $(-\infty, \infty)$; Range: $(0, \infty)$ D. Domain: $(-\infty, \infty)$; Range: $[0, \infty)$

E. Domain: $(0, \infty)$; Range: $(-\infty, \infty)$

3-7. Which of these functions are even: $f(x) = 3x^3$, $g(x) = 5x^2$, $h(x) = 2|x^3|$?

A. $f(x)$ only

B. $g(x)$ only

C. $g(x)$ and $h(x)$ only

D. $h(x)$ only

E. all three

3-8. Determine the symmetry of $f(x) = 3x^3 + 7x^5$.

A. Symmetric about the x-axis

B. Symmetric about the y-axis

C. Symmetric about the line $y = x$

D. Symmetric about the line $y = -x$

E. Symmetric about the origin

3-9. The graph of $g(x) = -9(x+8)^2 - 6$ is the graph of $f(x) = x^2$ with what transformations?

A. Horizontal shift of 8 units to the left, reflection about the y-axis, stretching by a factor of 9, and vertical shift of 6 units up

B. Horizontal shift of 8 units to the right, reflection about the y-axis, stretching by a factor of 9, and vertical shift of 6 units down

C. Horizontal shift of 8 units to the right, reflection about the x-axis, stretching by a factor of 9, and vertical shift of 6 units down

D. Horizontal shift of 8 units to the left, reflection about the x-axis, stretching by a factor of 9, and vertical shift of 6 units down

E. Horizontal shift of 8 units to the left, reflection about the y-axis, stretching by a factor of 9, and vertical shift of 6 units down

3-10. Rewrite the quadratic function $g(x) = x^2 - 3x - 4$ in a form that shows the shifts.

A. $\left(x - \frac{3}{2}\right)^2 - \frac{25}{4}$ B. $(x+3)^2 - \frac{25}{4}$ C. $(x-3)^2 - 9$

D. $\left(x - \frac{3}{2}\right)^2 + \frac{25}{4}$ E. $\left(x + \frac{3}{2}\right)^2 + \frac{25}{4}$

3-11. The graph shown here is the graph of

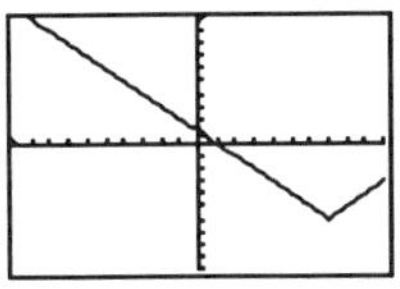

A. $f(x) = |x+7| - 6$ B. $f(x) = |x-7| - 6$ C. $f(x) = |x+6| + 7$

D. $f(x) = |x-6| - 7$ E. $f(x) = |x-7| + 6$

3-12 Determine the domain of $f(x) = \frac{(2x+3)(x-3)}{x-3}$.

A. All real numbers B. All real numbers except 3

C. All real numbers except $\frac{5}{2}$ D. All real numbers except –3

E. All real numbers except $-\frac{5}{2}$

3-13. Give the equation for all vertical asymptotes of $f(x) = \frac{x-3}{4x+7}$.

A. $x = \frac{7}{4}$ B. $x = 3$ C. $x = -\frac{7}{4}$ D. $y = \frac{1}{3}$ E. $y = -\frac{1}{3}$

3-14. Give the equation for all horizontal asymptotes of $f(x)=\dfrac{x-5}{2x+9}$.

A. $x=-\frac{9}{2}$ B. $x=5$ C. $y=\frac{1}{2}$ D. $y=-\frac{1}{2}$ E. $x=\frac{9}{2}$

3-15. Give the equation for all vertical asymptotes of the function graphed here.

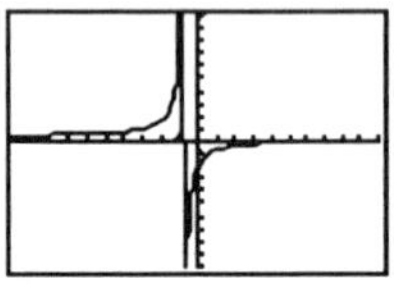

A. $x=-\frac{1}{4}$ B. $x=6$ C. $x=-\frac{3}{4}$ D. $x=\frac{1}{4}$ E. $y=\frac{3}{4}$

3-16. Give the equations of the asymptotes of $f(x)=x-2+\dfrac{3}{x+6}$.

A. Slant asymptote $y=x-2$

B. Slant asymptote $y=x-2$; horizontal asymptote $y=3$

C. Horizontal asymptote $y=2$

D. Slant asymptote $y=x-2$; vertical asymptote $x=-6$

E. Vertical asymptote $x=-6$

3-17. If $f(x)=\sqrt{x-3}$ and $g(x)=x+8$, write the expression for $(f+g)(x)$.

A. $\sqrt{x-3}+x+8$

B. $\sqrt{(2x+5)x}$

C. $\sqrt{2x-5}$

D. $\sqrt{2x+5}$

E. $\sqrt{x}+x$

3-18. $f(x)=x^2+1$ and $g(x)=\frac{-3}{x}$. Find $(g\circ f)(x)$.

A. $\frac{9}{x^2}+1$ B. $\frac{-3}{x^2+1}$ C. $\frac{3}{x^2+1}$

D. $\frac{3(x^2+1)}{x}$ E. $x^2+\frac{9}{x}$

3-19. A company has a total cost function of a commodity over a year given by $C(x)=\frac{x^3}{100}+5x^2+5000$ and a revenue function given by $R(x)=+8x^2+5000x$. If the profit function is $P(x)=R(x)-C(x)$, find $P(x)$.

A. $-2x^2+5000x$

B. $\frac{x^3}{100}-7x^2+5000x+5000$

C. $-\frac{x^3}{100}-7x^2+5000x-5000$

D. $\frac{x^3}{100}+3x^2+5000x+5000$

E. $-\frac{x^3}{100}+3x^2+5000x-5000$

3-20. $f(x)=x^3+1$; $g(x)=\sqrt{x+2}$. Find $(f\circ g)(2)$.

A. 18 B. 3.16 C. 9 D. 8 E. 3.32

3-21. Of $f(x)=5x^2$, $g(x)=3x^3$, and $h(x)=\sqrt{x-2}$, list all one-to-one functions.

A. $f(x)$ B. $g(x)$

C. $h(x)$ D. $g(x)$ and $h(x)$

E. $f(x), g(x)$, and $h(x)$

3-22. Of $f(x), g(x)$ and $h(x)$, list all one-to-one functions.

$f(x)$ 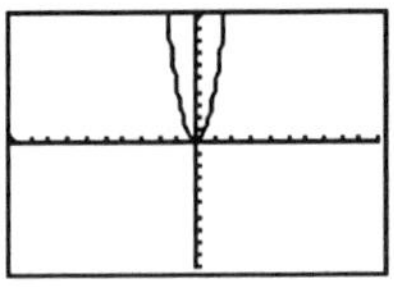

$g(x)$ 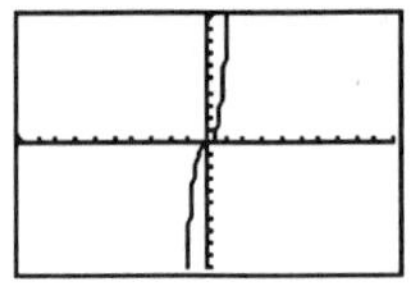

$h(x)$ 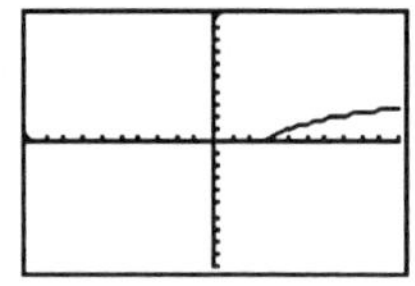

A. $f(x)$

B. $g(x)$

C. $h(x)$

D. $g(x)$ and $h(x)$

E. $f(x), g(x)$, and $h(x)$

3-23. Find the inverse function for $f(x)=7x-4$.

A. $f^{-1}(x)=\dfrac{1}{7x+4}$

B. $f^{-1}(x)=\dfrac{x+4}{7}$

C. $f^{-1}(x)=\dfrac{1}{7x}-4$

D. $f^{-1}(x)=\dfrac{1}{7x}+4$

E. $f^{-1}(x)=\dfrac{x}{7}+4$

3-24. Find the inverse function for $f(x)=x^3+5$.

A. $f^{-1}(x)=\dfrac{1}{x^3+5}$

B. $f^{-1}(x)=\dfrac{1}{x^3}+5$

C. $f^{-1}(x)=x^3+\dfrac{1}{5}$

D. $f^{-1}(x)=\sqrt[3]{x+5}$

E. $f^{-1}(x)=\sqrt[3]{x-5}$

3-25. If the velocity of a particle is $v(t)=\dfrac{t^3}{2}$, what is the expression for t as a function of v?

A. $\sqrt[3]{2v}$ B. $\sqrt[3]{2t}$ C. $\dfrac{2}{t^3}$ D. $2\sqrt[3]{v}$ E. $2\sqrt[3]{t}$

Test form 3D　　Date __________　　Name ______________________________

Course _______ Sec _______　　Student Number ____________________

Chapter 3

3-1. Determine the x-intercepts of the graph of the polynomial function $f(x) = x(x-2)(x+9)$

3-2. Find the polynomial function that has x-intercepts -3, 8, and 10, and y-intercept 480.

3-3. Determine where the quadratic function $f(x) = -x^2 + 13x - 30$ is increasing.

3-4. find the local minimum for $f(x) = x^2 + 15x + 50$.

3-5. Determine the local maximum for $f(x) = 2x^3 + 4x^2 - 6x + 6$ to four decimal places.

3-6. Determine the domain and range of $f(x) = x^2 - 10x + 25$.

3-7. Which of these functions are even: $f(x) = 10x^3$, $g(x) = 8x^2$, $h(x) = 6|x^3|$?

3-8. Determine the symmetry of $f(x) = 3x^3 + 10x^5$.

3-9. The graph of $g(x) = -7(x+9)^2 - 8$ is the graph of $f(x) = x^2$ with what transformations?

3-10. Rewrite the quadratic function $g(x) = x^2 - 2x - 9$ in a form that shows the shifts.

3-11. The graph shown here is the graph of

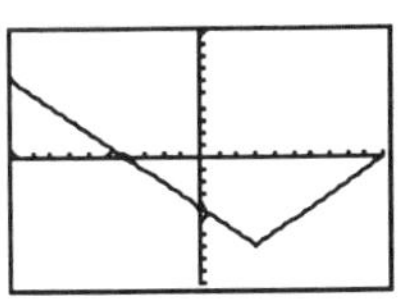

3-12. Determine the domain of $f(x) = \dfrac{(3x+5)(x-2)}{x-2}$.

3-13. Give the equation for all vertical asymptotes of $f(x) = \dfrac{x-4}{4x+9}$.

3-14. Give the equation for all horizontal asymptotes of $f(x) = \dfrac{x-2}{7x+3}$.

3-15. Give the equation for all vertical asymptotes of the function graphed here.

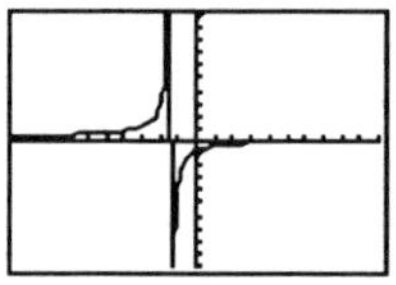

3-16. Give the equations of the asymptotes of $f(x) = x - 5 + \dfrac{7}{x+2}$.

3-17 If $f(x) = \sqrt{x-7}$ and $g(x) = x+2$, write the expression for $(f+g)(x)$.

3-18 $f(x) = x^2 + 28$ and $g(x) = \dfrac{-3}{x}$. Find $(g \circ f)(x)$.

3-19 A company has a total cost function of a commodity over a year given by $C(x) = \dfrac{x^3}{100} + x^2 + 5000$ and a revenue function given by $R(x) = +4x^2 + 5000x$. If the profit function is $P(x) = R(x) - C(x)$, find $P(x)$.

3-20 $f(x) = x^3 + 1$; $g(x) = \sqrt{x+2}$. Find $(f \circ g)(47)$.

3-21 Of $f(x) = 7x^2$, $g(x) = 3x^7$, and $h(x) = \sqrt{3x-2}$, list all one-to-one functions.

3-22 Of $f(x), g(x)$ and $h(x)$, list all one-to-one functions.

$f(x)$

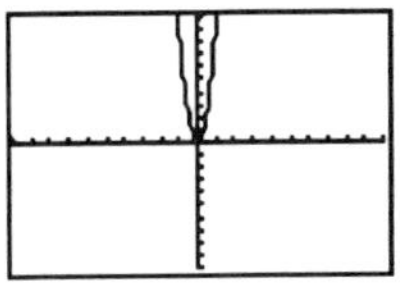

$g(x)$

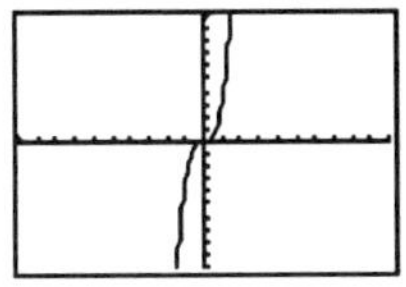

$h(x)$

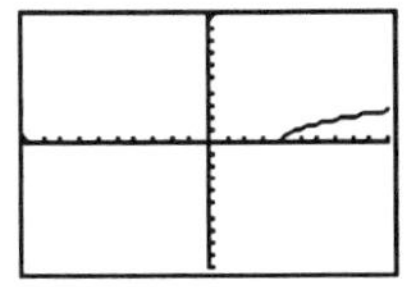

3-23 Find the inverse function for $f(x) = 6x - 25$.

3-24 Find the inverse function for $f(x) = x^7 + 8$.

3-25 If the velocity of a particle is $v(t) = \frac{t^5}{7}$, what is the expression for t as a function of v?

Test form 3E Date __________ Name ____________________________
Course _______ Sec _______ Student Number __________________

Chapter 3

3-1. Determine the x-intercepts of the graph of the polynomial function $f(x) = x(x-5)(x+12)$

3-2. Find the polynomial function that has x-intercepts –3, 8, and 10, and y-intercept 960.

3-3. Determine where the quadratic function $f(x) = -x^2 + 17x - 30$ is increasing.

3-4. Find the local minimum for $f(x) = x^2 + 5x + 20$.

3-5. Determine the local maximum for $f(x) = 5x^3 + 3x^2 - 6x + 6$ to four decimal places.

3-6. Determine the domain and range of $f(x) = x^2 - 14x + 49$.

3-7. Which of these functions are even: $f(x) = 10x^7$, $g(x) = 8x^{10}$, $h(x) = 16|x^7|$?

3-8. Determine the symmetry of $f(x) = 3x^5 + 10x^9$.

3-9. The graph of $g(x) = -3(x+10)^2 - 8$ is the graph of $f(x) = x^2$ with what transformations?

3-10. Rewrite the quadratic function $g(x) = x^2 - 2x - 19$ in a form that shows the shifts.

3-11. The graph shown here is the graph of

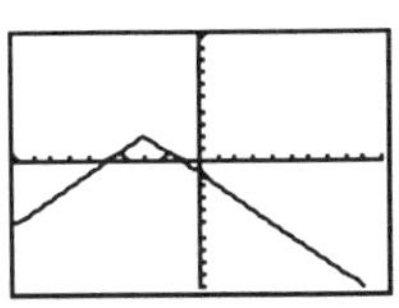

3-12. Determine the domain of $f(x) = \dfrac{(3x+6)(x-7)}{x-7}$.

3-13. Give the equation for all vertical asymptotes of $f(x) = \dfrac{x-4}{2x-3}$.

3-14. Give the equation for all horizontal asymptotes of $f(x) = \dfrac{6x-2}{7x+3}$.

3-15. Give the equation for all vertical asymptotes of the function graphed here.

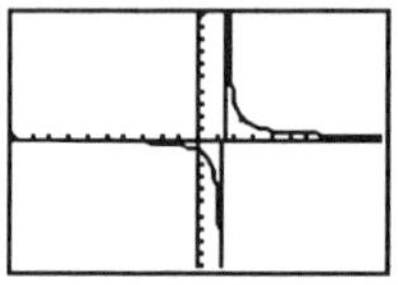

3-16. Give the equations of the asymptotes of $f(x) = x + 6 + \dfrac{7}{x+3}$.

3-17. If $f(x) = 2\sqrt{x-7}$ and $g(x) = x + 4$, write the expression for $(f+g)(x)$.

3-18. $f(x) = x^2 + 2$ and $g(x) = \dfrac{-10}{x}$. Find $(g \circ f)(x)$.

3-19. A company has a total cost function of a commodity over a year given by $C(x) = \dfrac{x^3}{100} + 11x^2 + 5000$ and a revenue function given by $R(x) = +14x^2 + 5000x$. If the profit function is $P(x) = R(x) - C(x)$, find $P(x)$.

3-20. $f(x) = x^3 + 1$; $g(x) = \sqrt{x+2}$. Find $(f \circ g)(79)$.

3-21. Of $f(x) = 6x^3$, $g(x) = 5x^6$, and $h(x) = \sqrt{3x-2}$, list all one-to-one functions.

3-22. Of $f(x), g(x)$ and $h(x)$, list all one-to-one functions.

$f(x)$

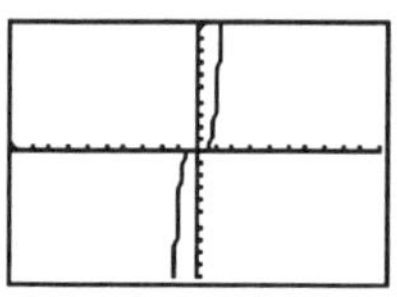

$g(x)$

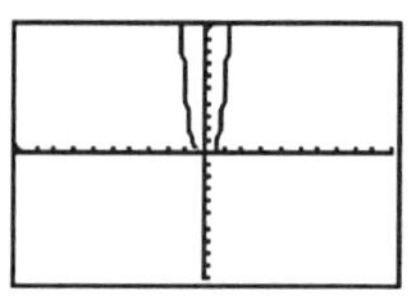

$h(x)$

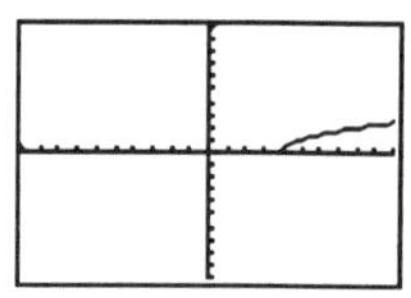

3-23. Find the inverse function for $f(x) = 10x + 7$.

3-24. Find the inverse function for $f(x) = x^3 + 26$.

3-25. If the velocity of a particle is $v(t) = \frac{t^{11}}{4}$, what is the expression for t as a function of v?

Test form 3F Date _________ Name _________________________
Course ______ Sec ______ Student Number _________________

Chapter 3

3-1. Determine the x-intercepts of the graph of the polynomial function $f(x) = x(x-7)(x+8)$

3-2. Find the polynomial function that has x-intercepts –3, 8, and 10, and y-intercept 720.

3-3. Determine where the quadratic function $f(x) = -x^2 + 15x - 60$ is increasing.

3-4. Find the local minimum for $f(x) = x^2 + 4x + 50$.

3-5. Determine the local maximum for $f(x) = 3x^3 + 2x^2 - 6x + 7$ to four decimal places.

3-6. Determine the domain and range of $f(x) = x^2 + 10x + 25$.

3-7. Which of these functions are even: $f(x) = 12x^5$, $g(x) = 4x^2$, $h(x) = 6|x^9|$?

3-8. Determine the symmetry of $f(x) = 3x^2 + 10x^4$.

3-9. The graph of $g(x) = -8(x+10)^2 - 7$ is the graph of $f(x) = x^2$ with what transformations?

3-10. Rewrite the quadratic function $g(x) = x^2 - 12x - 9$ in a form that shows the shifts.

3-11. The graph shown here is the graph of

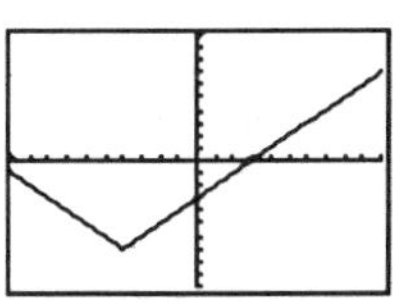

3-12. Determine the domain of $f(x) = \dfrac{(3x+7)(x-5)}{x-5}$.

3-13. Give the equation for all vertical asymptotes of $f(x) = \dfrac{x-4}{4x+15}$.

3-14. Give the equation for all horizontal asymptotes of $f(x) = \dfrac{3x-2}{7x+3}$.

3-15. Give the equation for all vertical asymptotes of the function graphed here.

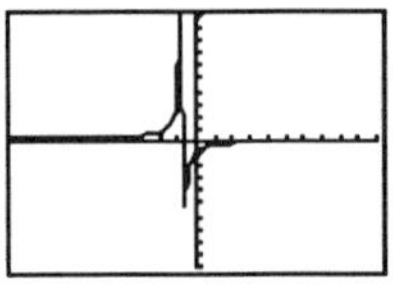

3-16. Give the equations of the asymptotes of $f(x) = x - 6 + \dfrac{7}{x+12}$.

3-17. If $f(x) = \sqrt{x-12}$ and $g(x) = x+5$, write the expression for $(f+g)(x)$.

3-18. $f(x) = x^2 + 16$ and $g(x) = \dfrac{3}{x}$. Find $(g \circ f)(x)$.

3-19. A company has a total cost function of a commodity over a year given by $C(x) = \dfrac{x^3}{100} - x^2 + 5000$ and a revenue function given by $R(x) = +2x^2 + 5000x$. If the profit function is $P(x) = R(x) - C(x)$, find $P(x)$.

3-20. $f(x) = x^3 + 1$; $g(x) = \sqrt{x+2}$. Find $(f \circ g)(62)$.

3-21. Of $f(x) = 7x^8$, $g(x) = 4x^3$, and $h(x) = \sqrt{5x-2}$, list all one-to-one functions.

3-22. Of $f(x), g(x)$ and $h(x)$, list all one-to-one functions.

$f(x)$

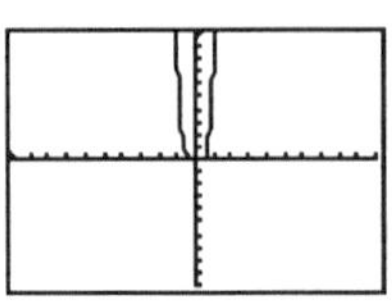

$g(x)$

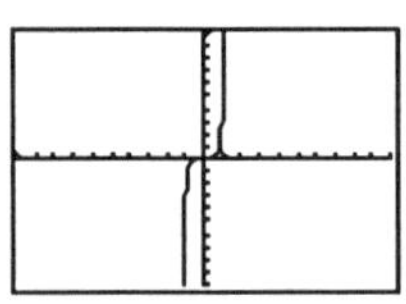

$h(x)$

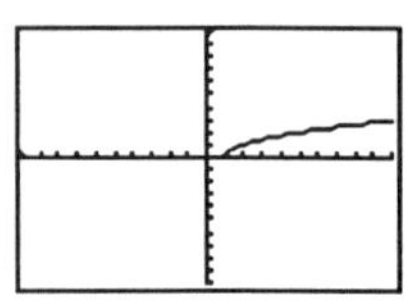

3-23. Find the inverse function for $f(x) = 3x - 21$.

3-24. Find the inverse function for $f(x) = x^9 + 23$.

3-25. If the velocity of a particle is $v(t) = \dfrac{t^9}{8}$, what is the expression for t as a function of v?

Test form 4A Date _________ Name _________________________
Course ______ Sec ______ Student Number _________________

Chapter 4

4-1. Use synthetic division to find the quotient and remainder when $f(x) = 2x^3 + 4x^2 + 5x + 1$ is divided by $g(x) = x + 2$.

A. $2x^2 + 5 + \frac{-9}{x+2}$

B. $2x^2 - 5 + \frac{-9}{x+2}$

C. $2x^2 - 5x - 9$; no remainder

D. $2x^2 - 5x + 9$; no remainder

E. $2x^2 - 5 + \frac{9}{x+2}$

4-2. Use synthetic division to find the quotient and remainder when the polynomial $f(x) = x^4 + 2x^2 + 5x + 7$ is divided by $g(x) = x - 2$.

A. $x^3 + 4x^2 + 13x + 33$; no remainder

B. $x^3 + 2x^2 + 6x + 17 + \frac{41}{x-2}$

C. $x^3 + 2x^2 + 6x + 17$; no remainder

D. $x^3 - 2x^2 + 6x - 17 + \frac{41}{x-2}$

E. $x^3 - 4x^2 + 13x - 33$

4-3. Use synthetic division with an appropriate divisor to compute $f(1)$ if $f(x) = 5x^3 + 2x^2 - 7x + 5$.

A. −5 B. 19 C. 5 D. −19 E. 0

4-4. Use synthetic division to determine a factor of $f(x) = x^3 - 2x^2 + 5x - 10$.

A. $x-2$ B. $x+2$ C. $x-4$

D. $x+4$ E. $x-\sqrt{2}$

4-5. Express the root $\sqrt{-9}$ in terms of i.

A. $9i$ B. $-9i$ C. $3i$ D. $-3i$ E. $-i$

4-6. Compute i^{31}.

A. 0 B. 1 C. -1 D. i E. $-i$

4-7. Find the complex conjugate of $5+3i$.

A. $5+3i$ B. $-5+3i$ C. $5-3i$

D. $-5-3i$ E. $15i$

4-8. Compute the product $(3+4i)(5-8i)$.

A. $47-4i$ B. $15+32i$ C. $15-32i$

D. $47i$ E. $17i$

4-9. Solve the equation $x^2 + 2x + 8 = 0$.

A. $-1 \pm i\sqrt{7}$ B. $1 \pm i\sqrt{7}$ C. $-1 \pm \sqrt{7}$

D. $1 \pm \sqrt{7}$ E. $-1 \pm 7i$

4-10. Perform the division $\dfrac{5+3i}{1+i}$.

A. 2 B. $2-2i$ C. $5+3i$

D. $4+i$ E. $4-i$

4-11. Compute the magnitude of 3 + 5i. Give the answer to four decimal places.

A. 8 B. 2 C. 8.4261

D. 4 E. 5.8310

4-12. Perform the additions and subtractions. Write the answer in the standard form $a+bi$. $(1+3i)+(2+2i)-(5+8i)$

A. $-2-3i$ B. $-2+3i$ C. $-2+12i$

D. $-4i$ E. -4

4-13. Use synthetic division to determine a factor of $f(x)=x^3+ix^2+4x+4i$.

A. $x+2i$ B. $x+4i$ C. $x-4i$

D. $x+3i$ E. $x-3i$

4-14. $f(x)=x^3+2x^2+(3+4i)x+5$ Use synthetic division to compute the value of $f(3)$.

A. 3 B. $71i$ C. $58+12i$

D. $60+12i$ E. $59+12i$

4-15. $f(x)=x^3+2ix^2+(1+2i)x+5$ Use synthetic division to compute the value of $f(2)$.

A. $15+15i$ B. $15+12i$ C. $16+14i$

D. $15+16i$ E. $16+16i$

4-16. Determine the zeros of the polynomial $f(x) = (x-4)^2(x+3)^3$ and state the multiplicity of each.

A. –4, multiplicity 2; 3 multiplicity 3

B. 4, multiplicity 2; –3 multiplicity 3

C. 2, multiplicity 4; 3 multiplicity –3

D. 2, multiplicity –4; 3 multiplicity 3

E. 4, multiplicity –2; 3 multiplicity –3

4-17. Determine the zeros of the polynomial $f(x) = (x+4i)^2(x-3+2i)^4$ and state the multiplicity of each.

A. $-4i$, multiplicity 2; $3-2i$ multiplicity 4

B. $4i$, multiplicity 2; $-3+2i$ multiplicity 4

C. $-4i$, multiplicity 2; 3.6056 multiplicity 4

D. 2, multiplicity 4; 4 multiplicity 3.6056

E. –2, multiplicity 4; –4 multiplicity 3.6056

4-18. One zero of $f(x) = x^3 - 3x^2 - 16x + 48$ is 3. Determine all of the zeros.

A. 3

B. −16, 16, 3

C. 3, 16 (multiplicity 2)

D. 3, 4 (multiplicity 2)

E. −4, 3, 4

4-19. Construct a polynomial $f(x)$ that has degree 3 with zeros –2, 1, and 2 such that $f(5) = 84$.

A. $x^3 - x^2 + 4x - 4$

B. $5x^3 - 5x^2 + 20x - 20$

C. $x^3 - x^2 - 4x + 4$

D. $5x^3 - 5x^2 - 20x + 20$

E. $x^3 - ix^2 + 4ix - 4$

4-20. Construct a polynomial $f(x)$ that has degree 3 with zeros $-2i$, 1, and $2i$ such that $f(5)=116$.

A. x^3-x^2+4x-4 B. $5x^3-5x^2+20x-20$

C. x^3-x^2-4x+4 D. $5x^3-5x^2-20x+20$

E. $x^3-ix^2+4ix-4$

4-21. Construct a polynomial $f(x)$ of degree 3 with real coefficients that has 4 and $3+i$ as zeros.

A. $2(x+4)(x+3+i)^2$ B. $2(x-4)(x+3+i)^2$

C. $2(x-4)(x-3-i)(x-3+i)$ D. $2(x+4)(x+3-i)(x+3+i)$

E. $2(x-4)(x-3-i)^2$

4-22. Find the real zeros of $f(x)=x^4+4x^3-22x^2-100x-75$.

A. $-5, -1, -3, 5$ B. $-1, -3$

C. $1, 3$ D. $-25, 1, 3, 25$

E. $-5, 1, 3, 5$

4-23. Find the real zeros of $f(x)=6x^4-x^3-97x^2+16x+16$.

A. $-\frac{1}{2}, \frac{1}{2}$ B. $-\frac{1}{3}, \frac{1}{2}$

C. $-4, -\frac{1}{3}, \frac{1}{2}, 4$ D. $-4, -\frac{1}{2}, \frac{1}{3}, 4$

E. $-16, -\frac{1}{3}, \frac{1}{2}, 16$

4-24. Find the real zeros of $f(x) = x^4 - 4x^3 - 2x^2 + 20x - 15$.

A. $-5, -1, -3, 5$ B. $-1, -3$

C. $1, 3$ D. $-\sqrt{5}, 1, 3, \sqrt{5}$

E. $-\sqrt{5}, -1, -3, \sqrt{5}$

4-25. Find the zeros of $f(x) = x^4 + 4x^3 + 7x^2 + 16x + 12$.

A. $-2i, -1, -3, 2i$ B. $-1, -3$

C. $1, 3$ D. $-4i, 1, 3, 4i$

E. $-4i, -1, -3, 4i$

Test form 4B Date _________ Name __________________________
Course ______ Sec ______ Student Number ________________

Chapter 4

4-1. Use synthetic division to find the quotient and remainder when $f(x) = 2x^3 + 4x^2 + 3x + 4$ is divided by $g(x) = x + 2$.

A. $2x^2 + 3 + \frac{-2}{x+2}$

B. $2x^2 - 3 + \frac{-2}{x+2}$

C. $2x^2 - 3x - 2$; no remainder

D. $2x^2 - 3x + 2$; no remainder

E. $2x^2 - 3 + \frac{2}{x+2}$

4-2. Use synthetic division to find the quotient and remainder when the polynomial $f(x) = x^4 + 2x^2 + 5x + 5$ is divided by $g(x) = x - 2$.

A. $x^3 + 4x^2 + 13x + 31$; no remainder

B. $x^3 + 2x^2 + 6x + 17 + \frac{39}{x-2}$

C. $x^3 + 2x^2 + 6x + 17$; no remainder

D. $x^3 - 2x^2 + 6x - 17 + \frac{34}{x-2}$

E. $x^3 - 4x^2 + 13x + 31$

4-3. Use synthetic division with an appropriate divisor to compute $f(1)$ if $f(x) = 5x^3 + 2x^2 - 7x + 7$.

A. −7 B. 7 C. 21 D. −21 E. 0

4-4. Use synthetic division to determine a factor of $f(x) = x^3 - 7x^2 + 5x - 35$.

A. $x + 7$ B. $x + 49$ C. $x - 49$

D. $x-7$ E. $x-\sqrt{7}$

4-5. Express the root $\sqrt{-16}$ in terms of i.

A. $4i$ B. $-4i$ C. $16i$ D. $-16i$ E. $-i$

4-6. Compute i^{35}.

A. 0 B. 1 C. -1 D. i E. $-i$

4-7. Find the complex conjugate of $4+7i$.

A. $4+7i$ B. $-4+7i$ C. $4-7i$

D. $-4-7i$ E. $28i$

4-8. Compute the product $(5+3i)(7-2i)$.

A. $37-6i$ B. $37+6i$ C. $43i$

D. $41+11i$ E. $31i$

4-9. Solve the equation $x^2+2x+11=0$.

A. $-1\pm i\sqrt{10}$ B. $1\pm i\sqrt{10}$ C. $-1\pm\sqrt{10}$

D. $1\pm\sqrt{10}$ E. $-1\pm 10i$

4-10. Perform the division $\dfrac{7+3i}{1+i}$.

A. 4 B. $4-4i$ C. $7+3i$

D. $5-2i$ E. $5+2i$

4-11. Compute the magnitude of $7 + 2i$. Give the answer to four decimal places.

A. 6.7082 B. 7.2801 C. 9

D. 5 E. 53

4-12. Perform the additions and subtractions. Write the answer in the standard form $a+bi$. $(1+3i)+(3+3i)-(5+7i)$

A. $1+3i$ B. $-1-i$ C. $-1+13i$

D. $2i$ E. 2

4-13. Use synthetic division to determine a factor of $f(x)=x^3+ix^2+9x+9i$.

A. $x+3i$ B. $x+9i$ C. $x-9i$

D. $x+4i$ E. $x-4i$

4-14. $f(x)=x^3+2x^2+(3+4i)x+5$ Use synthetic division to compute the value of $f(5)$.

A. 215 B. $215I$ C. $194+20i$

D. $195+20i$ E. $196+20i$

4-15. $f(x)=x^3+2ix^2+(1+2i)x+4$ Use synthetic division to compute the value of $f(4)$.

A. $14+14i$ B. $15+15i$ C. $15+14i$

D. $72+40i$ E. $73+40i$

4-16. Determine the zeros of the polynomial $f(x)=(x-5)^2(x+7)^3$ and state the multiplicity of each.

A. –5, multiplicity 2; 7 multiplicity 3

B. 5, multiplicity 2; –7 multiplicity 3

C. 2, multiplicity 5; 3 multiplicity –7

D. 2, multiplicity –5; 3 multiplicity 7

E. 5, multiplicity –2; 7 multiplicity –3

4-17. Determine the zeros of the polynomial $f(x)=(x+5i)^2(x-6+3i)^4$ and state the multiplicity of each.

A. $-5i$, multiplicity 2; $6-3i$ multiplicity 4

B. $5i$, multiplicity 2; $-6+3i$ multiplicity 4

C. $-5i$, multiplicity 2; 5 multiplicity 4

D. 2, multiplicity 5; 4 multiplicity 6.7082

E. –2, multiplicity 5; –4 multiplicity 6.7082

4-18. One zero of $f(x)=x^3-3x^2-81x+243$ is 3. Determine all of the zeros.

A. 3

B. –9, 3, 9

C. –81, 3, 81

D. 3, 9 (multiplicity 2)

E. 3, 81 (multiplicity 2)

4-19. Construct a polynomial $f(x)$ that has degree 3 with zeros –2, 1, and 2 such that $f(7)=270$.

A. x^3-x^2+4x-4

B. $7x^3-7x^2+28x-28$

C. x^3-x^2-4x+4

D. $7x^3-7x^2-28x+28$

E. $x^3-ix^2+4ix-4$

4-20. Construct a polynomial $f(x)$ that has degree 3 with zeros $-2i$, 1, and $2i$ such that $f(7)=318$.

A. $x^3 - x^2 + 4x - 4$

B. $7x^3 - 7x^2 + 28x - 28$

C. $x^3 - x^2 - 4x + 4$

D. $7x^3 - 7x^2 - 28x + 28$

E. $x^3 - ix^2 + 4ix - 4$

4-21. Construct a polynomial $f(x)$ of degree 3 with real coefficients that has 5 and $2 + i$ as zeros.

A. $2(x+5)(x+2+i)^2$

B. $2(x-5)(x+2+i)^2$

C. $2(x-5)(x-2-i)(x-2+i)$

D. $2(x+5)(x+2-i)(x+2+i)$

E. $2(x-5)(x-2-i)^2$

4-22. Find the real zeros of $f(x) = x^4 - 4x^3 - 33x^2 + 144x - 108$.

A. $-6, -1, -3, 6$

B. $-1, -3$

C. $1, 3$

D. $-36, 1, 3, 36$

E. $-6, 1, 3, 6$

4-23. Find the real zeros of $f(x) = 6x^4 - x^3 - 151x^2 + 25x + 25$.

A. $-\frac{1}{2}, \frac{1}{2}$

B. $-\frac{1}{3}, \frac{1}{2}$

C. $-5, -\frac{1}{3}, \frac{1}{2}, 5$

D. $-5, -\frac{1}{2}, \frac{1}{3}, 5$

E. $-25, -\frac{1}{3}, \frac{1}{2}, 25$

4-24. Find the real zeros of $f(x) = x^4 - 4x^3 + 12x - 9$.

A. $-9, -1, -3, 9$
B. $-1, -3$
C. $1, 3$
D. $-\sqrt{3}, 1, 3, \sqrt{3}$
E. $-1, -3$ (multiplicity 2), 3

4-25. Find the zeros of $f(x) = x^4 - 4x^3 + 12x^2 - 36x + 27$.

A. $-3i, -1, -3, 3i$
B. $-1, -3$
C. $1, 3$
D. $-9i, 1, 3, 9i$
E. $-3i, 1, 3, 3i$

Test form 4C   Date ________   Name ________________________
Course ______ Sec ______   Student Number ______________

Chapter 4

4-1. Use synthetic division to find the quotient and remainder when $f(x) = 2x^3 + 4x^2 + 9x + 3$ is divided by $g(x) = x + 2$.

A. $2x^2 - 9x - 15$; no remainder

B. $2x^2 - 9x + 15$; no remainder

C. $2x^2 - 9 + \frac{-15}{x+2}$

D. $2x^2 + 9 + \frac{-15}{x+2}$

E. $2x^2 - 9 + \frac{15}{x+2}$

4-2. Use synthetic division to find the quotient and remainder when the polynomial $f(x) = x^4 + 2x^2 + 5x + 2$ is divided by $g(x) = x - 2$.

A. $x^3 + 4x^2 + 13x + 28$; no remainder

B. $x^3 + 2x^2 + 6x + 17 + \frac{36}{x-2}$

C. $x^3 + 2x^2 + 6x + 17$; no remainder

D. $x^3 - 2x^2 + 6x - 17 + \frac{36}{x-2}$

E. $x^3 - 4x^2 + 13x + 28$

4-3. Use synthetic division with an appropriate divisor to compute $f(1)$ if $f(x) = 5x^3 + 2x^2 - 7x + 4$.

A. −4 B. 4 C. 18 D. −18 E. 0

4-4. Use synthetic division to determine a factor of $f(x) = x^3 - 6x^2 + 2x - 12$.

A. $x+6$ B. $x-6$ C. $x+36$

D. $x-36$ E. $x-\sqrt{6}$

4-5. Express the root $\sqrt{-81}$ in terms of i.

A. $81i$ B. $-81i$ C. $-9i$

D. $-i$ E. $9i$

4-6. Compute i^{43}.

A. 0 B. 1 C. -1 D. i E. $-i$

4-7. Find the complex conjugate of $2+9i$.

A. $2+9i$ B. $-2+9i$ C. $2-9i$

D. $-2-9i$ E. $18i$

4-8. Compute the product $(7+2i)(8-3i)$.

A. $56-6i$ B. $62-5i$ C. $56+6i$

D. $50i$ E. $62i$

4-9. Solve the equation $x^2 + 2x + 15 = 0$.

A. $1 \pm i\sqrt{14}$ B. $-1 \pm i\sqrt{14}$ C. $-1 \pm \sqrt{14}$

D. $1 \pm \sqrt{14}$ E. $-1 \pm 14i$

4-10. Perform the division $\dfrac{3+6i}{1+i}$.

A. -3 B. $-3+3i$ C. $3+6i$

D. $\dfrac{9+3i}{2}$ E. $\dfrac{9-3i}{2}$

4-11. Compute the magnitude of $6 + 7i$. Give the answer to four decimal places.

A. 3.6056 B. 1 C. 9.2195

D. 13 E. 14

4-12. Perform the additions and subtractions. Write the answer in the standard form $a+bi$. $(1+4i)+(3+2i)-(6+4i)$

A. $-2-2i$ B. $-2+10i$ C. $-4i$

D. 4 E. $-2+2i$

4-13. Use synthetic division to determine a factor of $f(x)=x^3+ix^2+16x+16i$.

A. $x+16i$ B. $x-16i$ C. $x+4i$

D. $x+5i$ E. $x-5i$

4-14. $f(x)=x^3+2x^2+(3+4i)x+5$ Use synthetic division to compute the value of $f(7)$.

A. $466+28i$ B. $467+28i$ C. $467+29i$

D. 495 E. $495i$

4-15. $f(x)=x^3+2ix^2+(1+2i)x+7$ Use synthetic division to compute the value of $f(2)$.

A. $17+12i$ B. $17+15i$ C. $16+14i$

D. $16+16i$ E. $15+15i$

4-16. Determine the zeros of the polynomial $f(x)=(x-6)^2(x+3)^3$ and state the multiplicity of each.

A. –6, multiplicity 2; 3 multiplicity 3

B. 6, multiplicity 2; –3 multiplicity 3

C. 2, multiplicity 6; 3 multiplicity –b

D. 2, multiplicity –6; 3 multiplicity b

E. 6, multiplicity –2; 3 multiplicity –3

4-17. Determine the zeros of the polynomial $f(x)=(x+7i)^2(x-6+3i)^4$ and state the multiplicity of each.

A. –7i, multiplicity 2; $6-3i$ multiplicity 4

B. 7i, multiplicity 2; –6 + 3i multiplicity 4

C. –7i, multiplicity 2; 6.7082 multiplicity 4

D. 2, multiplicity 7; 4 multiplicity 6.7082

E. –2, multiplicity 7; –4 multiplicity 6.7082

4-18. One zero of $f(x)=x^3-3x^2-100x+300$ is 3. Determine all of the zeros.

A. $-10,\ 3,\ 10$

B. 3

C. $-100,\ 3,\ 100$

D. 3, 10 (multiplicity 2)

E. 3, 100 (multiplicity 2)

4-19. Construct a polynomial $f(x)$ that has degree 3 with zeros –2, 1, and 2 such that $f(6)=160$.

A. x^3-x^2+4x-4

B. $6x^3-6x^2+24x-24$

C. x^3-x^2-4x+4

D. $6x^3-6x^2-24x+24$

E. $x^3-ix^2+4ix-4$

4-20. Construct a polynomial $f(x)$ that has degree 3 with zeros $-2i$, 1, and $2i$ such that $f(6) = 200$.

A. $x^3 - x^2 + 4x - 4$ B. $6x^3 - 6x^2 + 24x - 24$

C. $x^3 - x^2 - 4x + 4$ D. $6x^3 - 6x^2 - 24x + 24$

E. $x^3 - ix^2 + 4ix - 4$

4-21. Construct a polynomial $f(x)$ of degree 3 with real coefficients that has 8 and $3 + 2i$ as zeros.

A. $2(x+8)(x+3+2i)^2$

B. $2(x-8)(x+3+2i)^2$

C. $2(x-8)(x-3-2i)(x-3+2i)$

D. $2(x+8)(x+3-2i)(x+3+2i)$

E. $2(x-8)(x-3-2i)^2$

4-22. Find the real zeros of $f(x) = x^4 - 4x^3 - 13x^2 + 64x - 48$.

A. $-4, -1, -3, 4$ B. $-1, -3$

C. $1, 3$ D. $-16, 1, 3, 16$

E. $-4, 1, 3, 4$

4-23. Find the real zeros of $f(x) = 6x^4 - x^3 - 601x^2 + 100x + 100$.

A. $-\frac{1}{2}, \frac{1}{2}$ B. $-\frac{1}{3}, \frac{1}{2}$

C. $-10, -\frac{1}{3}, \frac{1}{2}, 10$ D. $-10, -\frac{1}{2}, \frac{1}{3}, 10$

E. $-100, -\frac{1}{3}, \frac{1}{2}, 100$

4-24. Find the real zeros of $f(x) = x^4 - 4x^3 - 7x^2 + 40x - 30$.

A. $-10, -1, -3, 10$ B. $-1, -3$

C. $1, 3$ D. $-\sqrt{10}, 1, 3, \sqrt{10}$

E. $-\sqrt{10}, -1, -3, \sqrt{10}$

4-25. Find the zeros of $f(x) = x^4 - 4x^3 + 103x^2 - 400x + 300$.

A. $-10i, -1, -3, 10i$ B. $-1, -3$

C. $1, 3$ D. $-100i, 1, 3, 100i$

E. $-10i, 1, 3, 10i$

Test form 4D | Date _________ | Name ___________________________
Course ______ Sec ______ | Student Number _______________

Chapter 4

4-1. Use synthetic division to find the quotient and remainder when $f(x) = 2x^3 + 4x^2 + 6x + 7$ is divided by $g(x) = x + 2$.

4-2. Use synthetic division to find the quotient and remainder when the polynomial $f(x) = x^4 + 2x^2 + 5x + 10$ is divided by $g(x) = x - 2$.

4-3. Use synthetic division with an appropriate divisor to compute $f(1)$ if $f(x) = 5x^3 + 2x^2 - 7x + 6$.

4-4. Use synthetic division to determine a factor of $f(x) = x^3 - 8x^2 + 3x - 24$.

4-5. Express the root $\sqrt{-36}$ in terms of i.

4-6. Compute i^{39}.

4-7. Find the complex conjugate of $5 + 2i$.

4-8. Compute the product $(9 + 2i)(3 - 4i)$.

4-9. Solve the equation $x^2 + 2x + 9 = 0$.

4-10. Perform the division $\frac{4 + 2i}{1 + i}$.

4-11. Compute the magnitude of $8 + 9i$. Give the answer to four decimal places.

4-12. Perform the additions and subtractions. Write the answer in the standard form $a + bi$. $(9 + 2i) + (3 + 4i) - (4 + 2i)$

4-13. Use synthetic division to determine a factor of $f(x) = x^3 + ix^2 + 25x + 25i$.

4-14. $f(x) = x^3 + 2x^2 + (3 + 4i)x + 5$ Use synthetic division to compute the value of $f(6)$.

4-15. $f(x) = x^3 + 2ix^2 + (1+2i)x + 8$ Use synthetic division to compute the value of $f(2)$.

4-16. Determine the zeros of the polynomial $f(x) = (x-7)^2(x+4)^3$ and state the multiplicity of each.

4-17. Determine the zeros of the polynomial $f(x) = (x+5i)^2(x-6+2i)^4$ and state the multiplicity of each.

4-18. One zero of $f(x) = x^3 - 3x^2 - 36x + 108$ is 3. Determine all of the zeros.

4-19. Construct a polynomial $f(x)$ that has degree 3 with zeros –2, 1, and 2 such that $f(9) = 616$.

4-20. Construct a polynomial $f(x)$ that has degree 3 with zeros $-2i$, 1, and $2i$ such that $f(9) = 680$.

4-21. Construct a polynomial $f(x)$ of degree 3 with real coefficients that has 7 and $5 + 3i$ as zeros.

4-22. Find the real zeros of $f(x) = x^4 - 4x^3 - 46x^2 + 196x - 147$.

4-23. Find the real zeros of $f(x) = 6x^4 + x^3 - 487x^2 - 81x + 81$.

4-24. Find the real zeros of $f(x) = x^4 - 4x^3 - 4x^2 + 28x - 21$.

4-25. Find the zeros of $f(x) = x^4 - 4x^3 + 39x^2 - 144x + 108$.

Test form 4E | Date ________ | Name ________________________
Course ______ Sec ______ | Student Number ______________

Chapter 4

4-1. Use synthetic division to find the quotient and remainder when $f(x) = 2x^3 + 4x^2 + 8x + 7$ is divided by $g(x) = x + 2$.

4-2. Use synthetic division to find the quotient and remainder when the polynomial $f(x) = x^4 + 2x^2 + 5x + 12$ is divided by $g(x) = x - 2$.

4-3. Use synthetic division with an appropriate divisor to compute $f(1)$ if $f(x) = 5x^3 + 2x^2 - 7x + 10$.

4-4. Use synthetic division to determine a factor of $f(x) = x^3 - 9x^2 + 2x - 18$.

4-5. Express the root $\sqrt{-121}$ in terms of i.

4-6. Compute i^{55}.

4-7. Find the complex conjugate of $7 + 3i$.

4-8. Compute the product $(4 + 3i)(8 - 5i)$.

4-9. Solve the equation $x^2 + 2x + 6 = 0$.

4-10. Perform the division $\dfrac{7 + 4i}{1 + i}$.

4-11. Compute the magnitude of $3 + 7i$. Give the answer to four decimal places.

4-12. Perform the additions and subtractions. Write the answer in the standard form $a + bi$. $(7 + 6i) + (2 + 3i) - (7 + 5i)$

4-13. Use synthetic division to determine a factor of $f(x) = x^3 + ix^2 + 36x + 36i$.

4-14. $f(x) = x^3 + 2x^2 + (3 + 4i)x + 5$ Use synthetic division to compute the value of $f(8)$.

4-15. $f(x) = x^3 + 2ix^2 + (1+2i)x + 18$ Use synthetic division to compute the value of $f(2)$.

4-16. Determine the zeros of the polynomial $f(x) = (x-8)^2(x+5)^3$ and state the multiplicity of each.

4-17. Determine the zeros of the polynomial $f(x) = (x+8i)^2(x-3+9i)^4$ and state the multiplicity of each.

4-18. One zero of $f(x) = x^3 - 3x^2 - 144x + 432$ is 3. Determine all of the zeros.

4-19. Construct a polynomial $f(x)$ that has degree 3 with zeros –2, 1, and 2 such that $f(10) = 864$.

4-20. Construct a polynomial $f(x)$ that has degree 3 with zeros $-2i$, 1, and $2i$ such that $f(10) = 936$.

4-21. Construct a polynomial $f(x)$ of degree 3 with real coefficients that has 4 and $5 + 2i$ as zeros.

4-22. Find the real zeros of $f(x) = x^4 - 4x^3 - 97x^2 + 400x - 300$.

4-23. Find the real zeros of $f(x) = 6x^4 + x^3 - 295x^2 - 49x + 49$.

4-24. Find the real zeros of $f(x) = x^4 - 4x^3 - 3x^2 + 24x - 18$.

4-25. Find the zeros of $f(x) = x^4 - 4x^3 + 28x^2 - 100x + 75$.

Test form 4F Date ________ Name ________________________
Course ______ Sec ______ Student Number ________________

Chapter 4

4-1. Use synthetic division to find the quotient and remainder when $f(x) = 2x^3 + 4x^2 + 3x + 9$ is divided by $g(x) = x + 2$.

4-2. Use synthetic division to find the quotient and remainder when the polynomial $f(x) = x^4 + 2x^2 + 5x + 19$ is divided by $g(x) = x - 2$.

4-3. Use synthetic division with an appropriate divisor to compute $f(1)$ if $f(x) = 5x^3 + 2x^2 - 7x + 15$.

4-4. Use synthetic division to determine a factor of $f(x) = x^3 - 2x^2 + 6x - 12$.

4-5. Express the root $\sqrt{-144}$ in terms of i.

4-6. Compute i^{61}.

4-7. Find the complex conjugate of $8 + 5i$.

4-8. Compute the product $(7 + 3i)(2 - 4i)$.

4-9. Solve the equation $x^2 + 2x + 7 = 0$.

4-10. Perform the division $\dfrac{13 + 3i}{1 + i}$.

4-11. Compute the magnitude of $4 + 7i$. Give the answer to four decimal places.

4-12. Perform the additions and subtractions. Write the answer in the standard form $a + bi$. $(5 + 8i) + (9 + 2i) - (6 + 3i)$

4-13. Use synthetic division to determine a factor of $f(x) = x^3 + ix^2 + 49x + 49i$.

4-14. $f(x) = x^3 + 2x^2 + (3 + 4i)x + 5$ Use synthetic division to compute the value of $f(4)$.

4-15. $f(x) = x^3 + 2ix^2 + (1+2i)x + 20$ Use synthetic division to compute the value of $f(2)$.

4-16. Determine the zeros of the polynomial $f(x) = (x-4)^2(x+8)^3$ and state the multiplicity of each.

4-17. Determine the zeros of the polynomial $f(x) = (x+7i)^2(x-8+2i)^4$ and state the multiplicity of each.

4-18. One zero of $f(x) = x^3 - 3x^2 - 64x + 192$ is 3. Determine all of the zeros.

4-19. Construct a polynomial $f(x)$ that has degree 3 with zeros –2, 1, and 2 such that $f(4) = 36$.

4-20. Construct a polynomial $f(x)$ that has degree 3 with zeros $-2i$, 1, and $2i$ such that $f(4) = 60$.

4-21. Construct a polynomial $f(x)$ of degree 3 with real coefficients that has 6 and $7 + 3i$ as zeros.

4-22. Find the real zeros of $f(x) = x^4 - 4x^3 - 78x^2 + 324x - 243$.

4-23. Find the real zeros of $f(x) = 6x^4 - x^3 - 55x^2 + 9x + 9$.

4-24. Find the real zeros of $f(x) = x^4 - 4x^3 - 8x^2 + 44x - 33$.

4-25. Find the zeros of $f(x) = x^4 - 4x^3 + 39x^2 - 144x + 108$.

Test form 5A Date ________ Name ______________________

Course ______ Sec ______ Student Number ________________

Chapter 5

5-1. Let $f(x) = 2^{-x}$. Compute $f(3)$ to four decimal places.

A. 0.1250 B. 8 C. −0.1250

D. −8 E. −9

5-2. Give the domain and range of 3^{x-2}.

A. Domain $(-\infty, \infty)$; Range $(0, \infty)$

B. Domain $(-\infty, \infty)$; Range $(2, \infty)$

C. Domain $(3, \infty)$; Range $(2, \infty)$

D. Domain $(-\infty, \infty)$; Range $(-2, \infty)$

E. Domain $(-\infty, \infty)$; Range $(-\infty, \infty)$

5-3. What are the shifts of the graph of a^x that are needed to produce the graph of $a^{x+2} + 12$?

A. Horizontal shift 2 units right, vertical shift 12 units up

B. Horizontal shift 2 units left, vertical shift 12 units down

C. Horizontal shift 2 units lcft, vcrtical shift 12 units up

D. Horizontal shift 12 units right, vertical shift 2 units down

E. Horizontal shift 12 units left, vertical shift 2 units up

5-4. The population of a certain city is described by the function $P(t) = 18(2^{t/5})$, where the population is in thousands and the time is in years. How long will it take for the population to reach 36?

A. 5 years B. 6 years C. 4 years

D. 3 years E. 7 years

5-5. Compute $e^{-2.3}$ to four decimal places.

A. −0.1003 B. 0.1003 C. −9.9742

D. 9.9742 E. 9.6223

5-6. Solve using algebraic methods. $5^{x-7} = 5^{2x}$

A. $-\frac{7}{3}$ B. $\frac{7}{3}$ C. 7

D. −7 E. No solution

5-7. Solve using algebraic methods. $2^{x-7} = 4^{x}$

A. $-\frac{7}{3}$ B. $\frac{7}{3}$ C. 7

D. −7 E. No solution

5-8. If $4^n = 64$, compute n^2.

A. 3 B. 5 C. 9 D. 4 E. 2

5-9. Solve the equation (if possible) to four decimal places. $4^{x+2} = 5$

A. −3.1918 B. 0.8390 C. −0.8390

D. 3.1610 E. No solution

5-10. Solve the equation (if possible) to four decimal places. $4^{x+2} = -12$

A. −3.1918 B. 0.8390 C. −0.8390

D. 3.1610 E. No solution

5-11. The sum of $7,000 is deposited in an account at the rate of 5% compounded monthly for 4 years. Compute the amount in the account at the end of the period.

A. $8508.54 B. $72898.89 C. $8546.27

D. $7117.40 E. $8549.82

5-12. The sum of $7,000 is deposited in an account at the rate of 5% compounded continuously for 4 years. Compute the amount in the account at the end of the period.

A. $8508.54 B. $72898.89 C. $8546.27

D. $7117.40 E. $8549.82

5-13. Find the inverse of $f(x) = 3^x$.

A. $\frac{1}{3^x}$ B. -3^x C. $\log_x 3$

D. $\log_3 x$ E. $-\frac{1}{3^x}$

5-14. Write $y = 3^x$ in logarithmic form.

A. $x = \log_y 3$ B. $x = \log_3 y$ C. $y = \log_b x$

D. $y = \log_x 3$ E. $y = x \log 3$

5-15. Write $y = \log_2 x$ in exponential form.

A. $y = 2^x$ B. $y = x^2$ C. $x = 2^y$

D. $x^y = 2$ E. $x = y^2$

5-16. Compute $\log_5 15625$.

A. 3 B. 4 C. 5 D. 6 E. 7

5-17. Determine ln 2.41 to four decimal places.

A. 0.3820 B. 0.8796 C. –0.3820

D. –0.8796 E. 11.1340

5-18. Find the inverse of $f(x) = \ln(x-2) + 3$.

A. $\dfrac{1}{\ln(x-2)+3}$ B. $\dfrac{1}{\ln(x-2)}+3$

C. $e^{x-3}+2$ D. $e^{x+3}-2$

E. Does not exist.

5-19. Find the inverse of $f(x) = e^{x+3} - 2$.

A. $\dfrac{1}{\ln(x-2)+3}$ B. $\dfrac{1}{\ln(x-2)}+3$

C. $e^{x-3}+2$ D. $e^{x+3}-2$

E. $\ln(x+2)-3$

5-20. Write the expression $\log_2(x^6 y^3 / z)$ in terms of logarithms of x, y, and z.

A. $6\log x + 3\log y - \log z$

B. $\log_2 x + \log_2 y - \log_2 z$

C. $6\log_2 x + 3\log_2 y - \log_2 z$

D. $\dfrac{\log_2 x^6 y^3}{\log_2 z}$

E. $\dfrac{(\log_2 x)^6 (\log_2 y)^3}{\log_2 z}$

5-21. Write the expression $\log_2 4x + \log_2(x-5) - \log_2(4(x-5))$ as a single logarithm.

A. $\log_2(x^4 - (x-5)^3)$

B. $\log_2\left(\dfrac{x^4}{(x-5)^{-3}}\right)$

C. $\log_2(-3(x-5)+x)$

D. $\log_2 x$

E. $\log_2\left(\dfrac{x^4}{(x-5)^3}\right)$

5-22. Use the properties of logarithms to solve the equation for *x*. $5\log_3 x - 4\log_3 x = 1$

A. 4 B. 3 C. 1 D. 0 E. No solution

5-23. Solve the equation (if possible) to four decimal places. $\ln(x^2 + 4x + 3) = x - 5$

A. −0.7952

B. 4.2048, 8.2986

C. 9.9553

D. 8.2986

E. 3.2986

5-24. Solve the equation for *x*, to four decimal places, by writing it in logarithmic form, then use the calculator. $5 = e^{3/x}$

A. 1.8640 B. 4.8283 C. 1.6667

D. −4.8283 E. −1.8640

5-25. Use exponential regression to arrive at an exponential equation from the data.

t	0	1	2	3	4	5
Value	3	12	48	192	768	3072

A. $3(4^x) - 3$ B. $\dfrac{4^x}{3}$ C. $(12)^x$

D. 4^x E. $3(4^x)$

Test form 5B Date ________ Name ________________________
Course ______ Sec ______ Student Number ________________

Chapter 5

5-1. Let $f(x) = 2^{-x}$. Compute $f(4)$ to four decimal places.

A. 16 B. 0.0625 C. −16

D. −0.0625 E. −15

5-2. Give the domain and range of 4^{x-5}.

A. Domain $(-\infty, \infty)$; Range $(0, \infty)$

B. Domain $(-\infty, \infty)$; Range $(5, \infty)$

C. Domain $(4, \infty)$; Range $(5, \infty)$

D. Domain $(-\infty, \infty)$; Range $(-5, \infty)$

E. Domain $(-\infty, \infty)$; Range $(-\infty, \infty)$

5-3. What are the shifts of the graph of a^x that are needed to produce the graph of $a^{x+5} + 15$?

A. Horizontal shift 5 units left, vertical shift 15 units up

B. Horizontal shift 5 units left, vertical shift 15 units down

C. Horizontal shift 5 units right, vertical shift 15 units up

D. Horizontal shift 15 units right, vertical shift 5 units up

E. Horizontal shift 15 units left, vertical shift 5 units down

5.4. The population of a certain city is described by the function $P(t) = 19(2^{t/5})$, where the population is in thousands and the time is in years. How long will it take for the population to reach 38?

A. 7 years B. 6 years C. 4 years

D. 3 years E. 5 years

5-5. Compute $e^{-4.1}$ to four decimal places.

A. 60.3403 B. –60.3403 C. –0.0166

D. 0.0166 E. 46.3147

5-6. Solve using algebraic methods. $7^{x-5} = 7^{2x}$

A. $-\frac{5}{3}$ B. $\frac{5}{3}$ C. -5

D. 5 E. No solution

5-7. Solve using algebraic methods. $6^{x-5} = 36^{x}$

A. $-\frac{5}{3}$ B. $\frac{5}{3}$ C. -5

D. 5 E. No solution

5-8. If $5^n = 125$, compute n^2.

A. 3 B. 5 C. 9 D. 4 E. 2

5-9. Solve the equation (if possible) to four decimal places. $7^{x+2} = 4$

A. 2.7124 B. −1.2876 C. –0.5963

D. 3.4037 E. No solution

5-10. Solve the equation (if possible) to four decimal places. $7^{x+2} = -4$

A. 2.7124 B. −1.2876 C. −0.5963

D. 3.4037 E. No solution

5-11. The sum of $9,000 is deposited in an account at the rate of 5% compounded monthly for 4 years. Compute the amount in the account at the end of the period.

A. $93611.43 B. $9150.94 C. $10939.56

D. $10988.06 E. $10992.62

5-12. The sum of $9,000 is deposited in an account at the rate of 5% compounded continuously for 4 years. Compute the amount in the account at the end of the period.

A. $93611.43 B. $9150.94 C. $10939.56

D. $10988.06 E. $10992.62

5-13. Find the inverse of $f(x) = 6^x$.

A. $\frac{1}{6^x}$ B. $\log_6 x$ C. $\log_x 6$

D. -6^x E. $-\frac{1}{6^x}$

5-14. Write $y = 7^x$ in logarithmic form.

A. $x = \log_y 7$ B. $y = \log_7 x$ C. $x = \log_7 y$

D. $y = \log_x 7$ E. $y = x \log 7$

5-15. Write $y = \log_5 x$ in exponential form.

A. $y = 5^x$ B. $x = 5^y$ C. $y = x^5$

D. $x^y = 5$ E. $x = y^5$

5-16. Compute $\log_4 4096$.

A. 3 B. 4 C. 5 D. 6 E. 7

5-17. Determine ln 2.05 to four decimal places.

A. 0.3118 B. −0.3118 C. 0.7178

D. −0.7178 E. 7.7679

5-18. Find the inverse of $f(x) = \ln(x-4) + 3$.

A. $\dfrac{1}{\ln(x-4)+3}$ B. $\dfrac{1}{\ln(x-4)}+3$

C. $e^{x-3}+4$ D. $e^{x+3}-4$

E. Does not exist.

5-19. Find the inverse of $f(x) = e^{x+3} - 4$.

A. $\dfrac{1}{\ln(x-4)+3}$ B. $\dfrac{1}{\ln(x-4)}+3$

C. $e^{x-3}+4$ D. $e^{x+3}-4$

E. $\ln(x+4)-3$

5-20. Write the expression $\log_2(x^5 y^3 / z)$ in terms of logarithms of x, y, and z.

A. $5\log x + 3\log y - \log z$

B. $\log_2 x + \log_2 y - \log_2 z$

C. $5\log_2 x + 3\log_2 y - \log_2 z$

D. $\dfrac{\log_2 x^5 y^3}{\log_2 z}$

E. $\dfrac{(\log_2 x)^5 (\log_2 y)^3}{\log_2 z}$

5-21. Write the expression $\log_5 6x + \log_5(x-3) - \log_5(6(x-3))$ as a single logarithm.

A. $\log_5 x$

B. $\log_5(x^6(x-3)^5)$

C. $\log_5(-5(x-3)+x)$

D. $\log_5\left(\dfrac{x^6}{(x-3)^{-5}}\right)$

E. $\log_5 \dfrac{x^6}{(x-3)^5}$

5-22. Use the properties of logarithms to solve the equation for x. $8\log_3 x - 7\log_3 x = 1$

A. 3 B. 7 C. 1 D. 0 E. No solution

5-23. Solve the equation (if possible) to four decimal places. $\ln(x^2+4x+3) = x-6$

A. −0.7952

B. 5.2948, 9.2986

C. 11.1465

D. 9.2986

E. 3.2986

5-24. Solve the equation for x, to four decimal places, by writing it in logarithmic form, then use the calculator. $6 = e^{5/x}$

A. 1.8640 B. 2.7906 C. 1.6667

D. −4.8283 E. −1.8640

5-25. Use exponential regression to arrive at an exponential equation from the data.

t	0	1	2	3	4	5
Value	3	15	75	375	1875	9375

A. $3(5^x)-3$ B. $\dfrac{5^x}{3}$ C. $(15)^x$

D. 5^x E. $3(5^x)$

Test form 5C　　Date ________　　Name ______________________
Course ______ Sec ______　　Student Number ________________

Chapter 5

5-1. Let $f(x) = 2^{-x}$. Compute $f(5)$ to four decimal places.

A. 32　　B. −32　　C. 0.0313

D. −0.0313　　E. 25

5-2. Give the domain and range of 6^{x-8}.

A. Domain $(-\infty, \infty)$; Range $(0, \infty)$

B. Domain $(-\infty, \infty)$; Range $(8, \infty)$

C. Domain $(6, \infty)$; Range $(8, \infty)$

D. Domain $(-\infty, \infty)$; Range $(-8, \infty)$

E. Domain $(-\infty, \infty)$; Range $(-\infty, \infty)$

5-3. What are the shifts of the graph of a^x that are needed to produce the graph of $a^{x+8} + 11$?

A. Horizontal shift 8 units right, vertical shift 11 units up

B. Horizontal shift 8 units left, vertical shift 11 units up

C. Horizontal shift 8 units left, vertical shift 11 units down

D. Horizontal shift 11 units right, vertical shift 8 units up

E. Horizontal shift 11 units left, vertical shift 8 units down

5-4. The population of a certain city is described by the function $P(t) = 15(2^{t/6})$, where the population is in thousands and the time is in years. How long will it take for the population to reach 30?

A. 7 years B. 6 years C. 5 years

D. 4 years E. 8 years

5-5. Compute $e^{-5.2}$ to four decimal places.

A. 0.0055 B. −0.0055 C. 181.2722

D. −181.2722 E. 88.3688

5-6. Solve using algebraic methods. $3^{x-8} = 3^{2x}$

A. $-\frac{8}{3}$ B. $\frac{8}{3}$ C. −8

D. 8 E. No solution

5-7. Solve using algebraic methods. $4^{x-8} = 16^x$

A. $-\frac{8}{3}$ B. $\frac{8}{3}$ C. −8

D. 8 E. No solution

5-8. If $2^n = 8$, compute n^2.

A. 3 B. 5 C. 9 D. 4 E. 2

5-9. Solve the equation (if possible) to four decimal places. $6^{x+2} = 11$

A. −0.6617 B. 3.3383 C. −0.5963

D. 2.7472 E. No solution

5-10. Solve the equation (if possible) to four decimal places. $6^{x+2} = -11$

A. −0.6617 B. 3.3383 C. −0.5963

D. 2.7472 E. No solution

5-11. The sum of $5,000 is deposited in an account at the rate of 5% compounded monthly for 4 years. Compute the amount in the account at the end of the period.

A. $6104.48 B. $6077.53 C. $52006.36

D. $5083.86 E. $6107.01

5-12. The sum of $5,000 is deposited in an account at the rate of 5% compounded continuously for 4 years. Compute the amount in the account at the end of the period.

A. $6104.48 B. $6077.53 C. $52006.36

D. $5083.86 E. $6107.01

5-13. Find the inverse of $f(x) = 7^x$.

A. $\log_x 7$ B. $\log_7 x$ C. $\dfrac{1}{7^x}$

D. -7^x E. $-\dfrac{1}{7^x}$

5-14. Write $y = 4^x$ in logarithmic form.

A. $x = \log_y 4$ B. $y = \log_4 x$ C. $y = \log_x 4$

D. $x = \log_4 y$ E. $y = x \log 4$

5-15. Write $y = \log_6 x$ in exponential form.

A. $x = 6^y$ B. $y = 6^x$ C. $y = x^6$

D. $x^y = 6$ E. $x = y^6$

5-16. Compute $\log_3 729$.

A. 3 B. 4 C. 5 D. 6 E. 7

5-17. Determine ln 2.15 to four decimal places.

A. 0.7655 B. 0.3324 C. −0.7655

D. −0.3324 E. 8.5849

5-18. Find the inverse of $f(x) = \ln(x-2) + 5$.

A. $\dfrac{1}{\ln(x-2)+5}$ B. $\dfrac{1}{\ln(x-2)}+5$

C. $e^{x-5}+2$ D. $e^{x+5}-2$

E. Does not exist.

5-19. Find the inverse of $f(x) = e^{x+5} - 2$.

A. $\dfrac{1}{\ln(x-2)+5}$ B. $\dfrac{1}{\ln(x-2)}+5$

C. $e^{x-5}+2$ D. $e^{x+5}-2$

E. $\ln(x+2)-5$

5-20. Write the expression $\log_2(x^4y^2/z)$ in terms of logarithms of x, y, and z.

A. $4\log x + 2\log y - \log z$

B. $\log_2 x + \log_2 y - \log_2 z$

C. $4\log_2 x + 2\log_2 y - \log_2 z$

D. $\dfrac{\log_2 x^4y^2}{\log_2 z}$

E. $\dfrac{(\log_2 x)^2(\log_2 y)^2}{\log_2 z}$

5-21. Write the expression $\log_3 7x + \log_3(x-2) - \log_3(7(x-2))$ as a single logarithm.

A. $\log_3(x^7 - (x-2)^6)$

B. $\log_3 x$

C. $\log_3(-6(x-2)+x)$

D. $\log_3\left(\frac{x^7}{(x-2)^{-6}}\right)$

E. $\log_3\left(\frac{x^7}{(x-2)^6}\right)$

5-22. Use the properties of logarithms to solve the equation for x. $3\log_7 x - 2\log_7 x = 1$

A. 6 B. 1 C. 0 D. 7 E. No solution

5-23. Solve the equation (if possible) to four decimal places. $\ln(x^2 + 4x + 3) = x - 9$

A. −0.7952

B. 8.2048, 12.3986

C. 14.6173

D. 12.3986

E. 3.2986

5-24. Solve the equation for x, to four decimal places, by writing it in logarithmic form, then use the calculator. $7 = e^{4/x}$

A. 7.7836

B. −7.7836

C. 1.75

D. −2.0556

E. 2.0556

5-25. Use exponential regression to arrive at an exponential equation from the data.

t	0	1	2	3	4	5
Value	3	21	147	1029	7203	50421

A. $3(7^x) - 3$

B. $\frac{7^x}{3}$

C. $(21)^x$

D. 7^x

E. $3(7^x)$

Test form 5D Date ________ Name ________________________

Course ______ Sec ______ Student Number ________________

Chapter 5

5-1. Let $f(x) = 2^{-x}$. Compute $f(6)$ to four decimal places.

5-2. Give the domain and range of 7^{x-1}.

5-3. What are the shifts of the graph of a^x that are needed to produce the graph of $a^{x+1} + 14$?

5.4. The population of a certain city is described by the function $P(t) = 17(2^{t/8})$, where the population is in thousands and the time is in years. How long will it take for the population to reach 34?

5-5. Compute $e^{-7.1}$ to four decimal places.

5-6. Solve using algebraic methods. $10^{x-8} = 10^{2x}$

5-7. Solve using algebraic methods. $5^{x-10} = 25^x$

5-8. If $6^n = 216$, compute n^2.

5-9. Solve the equation (if possible) to four decimal places. $8^{x+2} = 5$

5-10. Solve the equation (if possible) to four decimal places. $8^{x+2} = -5$

5-11. The sum of $12,000 is deposited in an account at the rate of 5% compounded monthly for 4 years. Compute the amount in the account at the end of the period.

5-12. The sum of $12,000 is deposited in an account at the rate of 5% compounded continuously for 4 years. Compute the amount in the account at the end of the period.

5-13. Find the inverse of $f(x) = 4^x$.

5-14. Write $y = 8^x$ in logarithmic form.

5-15. Write $y = \log_3 x$ in exponential form.

5-16. Compute $\log_5 625$

5-17. Determine ln 2.33 to four decimal places.

5-18. Find the inverse of $f(x) = \ln(x - 7) + 3$.

5-19. Find the inverse of $f(x) = e^{x+7} - 3$.

5-20. Write the expression $\log_3(x^7 y^3 / z)$ in terms of logarithms of x, y, and z.

5-21. Write the expression $\log_4 8x + \log_4(x-2) - \log_4(4(x-2))$ as a single logarithm.

5-22. Use the properties of logarithms to solve the equation for x. $5\log_8 x - 4\log_8 x = 1$

5-23. Solve the equation (if possible) to four decimal places. $\ln(x^2 + 4x + 3) = x - 4$

5-24. Solve the equation for x, to four decimal places, by writing it in logarithmic form, then use the calculator. $9 = e^{5/x}$

5-25. Use exponential regression to arrive at an exponential equation from the data.

t	0	1	2	3	4	5
Value	3	24	192	1536	12288	98308

Test form 5E Date ________ Name ______________________
Course ______ Sec ______ Student Number ______________

Chapter 5

5-1. Let $f(x) = 2^{-x}$. Compute $f(7)$ to four decimal places.

5-2. Give the domain and range of 8^{x-2}.

5-3. What are the shifts of the graph of a^x that are needed to produce the graph of $a^{x+7} + 12$?

5-4. The population of a certain city is described by the function $P(t) = 14(2^{t/7})$, where the population is in thousands and the time is in years. How long will it take for the population to reach 28?

5-5. Compute $e^{-7.4}$ to four decimal places.

5-6. Solve using algebraic methods. $4^{x-7} = 4^{2x}$

5-7. Solve using algebraic methods. $3^{x-2} = 9^x$

5-8. If $8^n = 512$, compute n^2.

5-9. Solve the equation (if possible) to four decimal places. $4^{x+2} = 9$

5-10. Solve the equation (if possible) to four decimal places. $4^{x+2} = -9$

5-11. The sum of $15,000 is deposited in an account at the rate of 5% compounded monthly for 4 years. Compute the amount in the account at the end of the period.

5-12. The sum of $15,000 is deposited in an account at the rate of 5% compounded continuously for 4 years. Compute the amount in the account at the end of the period.

5-13. Find the inverse of $f(x) = 8^x$.

5-14. Write $y = 2^x$ in logarithmic form.

5-15. Write $y = \log_9 x$ in exponential form.

5-16. Compute $\log_4 1024$

5-17. Determine ln 2.49 to four decimal places.

5-18. Find the inverse of $f(x) = \ln(x-9) + 4$.

5-19. Find the inverse of $f(x) = e^{x+9} - 4$

5-20. Write the expression $\log_5(x^5 y^4 / z)$ in terms of logarithms of x, y, and z.

5-21. Write the expression $\log_3 9x + \log_3(x-1) - \log_3(9(x-1))$ as a single logarithm.

5-22. Use the properties of logarithms to solve the equation for x. $8\log_7 x - 7\log_7 x = 1$

5-23. Solve the equation (if possible) to four decimal places. $\ln(x^2 + 4x + 3) = x - 7$

5-24. Solve the equation for x, to four decimal places, by writing it in logarithmic form, then use the calculator. $8 = e^{2/x}$

5-25. Use exponential regression to arrive at an exponential equation from the data.

t	0	1	2	3	4	5
Value	3	18	108	648	3888	23328

Test form 5F Date _________ Name ___________________________
Course ______ Sec ______ Student Number _________________

Chapter 5

5-1. Let $f(x) = 2^{-x}$. Compute $f(8)$ to four decimal places.

5-2. Give the domain and range of 5^{x-3}.

5-3. What are the shifts of the graph of a^x that are needed to produce the graph of $a^{x+9} + 19$?

5-4. The population of a certain city is described by the function $P(t) = 15(2^{t/9})$, where the population is in thousands and the time is in years. How long will it take for the population to reach 30?

5-5. Compute $e^{-6.5}$ to four decimal places.

5-6. Solve using algebraic methods. $3^{x-5} = 3^{2x}$

5-7. Solve using algebraic methods. $7^{x-5} = 49^x$

5-8. If $7^n = 343$, compute n^2.

5-9. Solve the equation (if possible) to four decimal places. $3^{x+2} = 8$

5-10. Solve the equation (if possible) to four decimal places. $3^{x+2} = -8$

5-11. The sum of $8,000 is deposited in an account at the rate of 5% compounded monthly for 4 years. Compute the amount in the account at the end of the period.

5-12. The sum of $8,000 is deposited in an account at the rate of 5% compounded continuously for 4 years. Compute the amount in the account at the end of the period.

5-13. Find the inverse of $f(x) = 9^x$.

5-14. Write $y=6^x$ in logarithmic form.

5-15. Write $y=\log_{11}x$ in exponential form.

5-16. Compute $\log_3 243$

5-17. Determine ln 2.68 to four decimal places.

5-18. Find the inverse of $f(x)=\ln(x-1)+3$.

5-19. Find the inverse of $f(x)=e^{x+1}-3$.

5-20. Write the expression $\log_6(x^7y^6/z)$ in terms of logarithms of x, y, and z.

5-21. Write the expression $\log_5 8x+\log_5(x-9)-\log_5(8(x-9))$ as a single logarithm.

5-22. Use the properties of logarithms to solve the equation for x. $9\log_3 x-8\log_3 x=1$

5-23. Solve the equation (if possible) to four decimal places. $\ln(x^2+4x+3)=x-8$

5-24. Solve the equation for x, to four decimal places, by writing it in logarithmic form, then use the calculator. $6=e^{9/x}$

5-25. Use exponential regression to arrive at an exponential equation from the data.

t	0	1	2	3	4	5
Value	3	6	12	24	48	96

Test form 6A Date ________ Name ____________________
Course ______ Sec ______ Student Number ______________

Chapter 6

6-1. Solve the system of equations using the method of substitution.
$2x+5y=12$
$x+y=3$

A. (0, 0) B. (2, 5) C. (−1, 4)

D. (2, 1) E. (1, 2)

6-2. Solve the system of equations using the method of substitution.
$2x+5y=12$
$x+3y=7$

A. (0, 0) B. (2, 5) C. (−1, 4)

D. (2, 1) E. (1, 2)

6-3. Solve the system of equations using the method of elimination.

$4x+3y=11$
$2x+3y=7$

A. (0, 0) B. (4, 3) C. (−1, 4)

D. (2, 1) E. (1, 2)

6-4. Solve the system of equations using the method of elimination.
$3x+4y=11$
$4x-3y=-2$

A. (0, 0) B. (1, 2) C. (2, 1)

D. (−1, 4) E. (3, 4)

6-5. Solve the system of equations using the method of elimination. Give answer to four decimal places.

$4x + 20y = 32$

$3x + y = 9$

A. (2.6429, 1.0714) B. (1.6429, 2.0714)

C. (3.0003, −0.0001) D. (0.4511, 8.5489)

E. (8.5489, 0.4511)

6-6. Solve the system of equations using the method of elimination. Give answer to four decimal places.

$2x + 4y = 8$

$-4x - 8y = 3$

A. (1, 3) B. (3, 1) C. (2, 1)

D. (4, 0) E. No solution

6-7. A person has 13 coins with a value of $2.35. If the coins are all quarters and dimes, how many of each does the person have?

A. 7 quarters, 8 dimes B. 7 quarters, 6 dimes

C. 10 quarters, 8 dimes D. 10 quarters, 6 dimes

E. 0 quarters, 13 dimes

6-8. Determine the augmented matrix for the system

$2x + 3y = 6$

$3x + 4y = 9$

A. $\begin{bmatrix} 2 & 3 & 6 \\ 3 & 4 & 9 \end{bmatrix}$ B. $\begin{bmatrix} 2 & 3 \\ 3 & 4 \\ 6 & 9 \end{bmatrix}$ C. $\begin{bmatrix} 2 & 3 \\ 3 & 4 \\ 6 & 9 \end{bmatrix}$

D. $\begin{bmatrix} 2 & 3 \\ 5 & 4 \end{bmatrix}$ E. $\begin{bmatrix} 5 & 6 \\ 9 & 9 \end{bmatrix}$

6-9. Interpret the matrix as an augmented matrix of a system of equations. Write down the system.

$$\begin{bmatrix} 2 & 4 & 7 \\ 3 & 5 & 9 \end{bmatrix}$$

A. $\begin{aligned} 2x+3y&=7 \\ 4x+5y&=9 \end{aligned}$

B. $\begin{aligned} 2x+4y&=7 \\ 3x+5y&=9 \end{aligned}$

C. $\begin{aligned} 2x+4y+7z&=0 \\ 3x+5y+9z&=0 \end{aligned}$

D. $\begin{aligned} 4x+7y&=2 \\ 5x+9y&=3 \end{aligned}$

E. $\begin{aligned} 2x+7y&=4 \\ 3x+9y&=5 \end{aligned}$

6-10. Starting with $\begin{bmatrix} 1 & 0 & 2 & 1 \\ 3 & 4 & 5 & 3 \\ 8 & 5 & 7 & 4 \end{bmatrix}$, perform the transformation $-3R_1 + R_2$.

A. $\begin{bmatrix} 1 & 0 & 2 & 1 \\ 6 & 4 & 11 & 5 \\ 8 & 5 & 7 & 4 \end{bmatrix}$

B. $\begin{bmatrix} 1 & 0 & 2 & 1 \\ 0 & 4 & -1 & 0 \\ 8 & 5 & 7 & 4 \end{bmatrix}$

C. $\begin{bmatrix} 1 & 0 & 2 & 1 \\ 0 & 4 & 0 & 2 \\ 8 & 5 & 7 & 4 \end{bmatrix}$

D. $\begin{bmatrix} 0 & 4 & -1 & 2 \\ 8 & 5 & 7 & 4 \end{bmatrix}$

E. $\begin{bmatrix} 0 & 4 & 0 & 2 \\ 8 & 5 & 7 & 4 \end{bmatrix}$

6-11. Solve the system of equations using the method of Gauss-Jordan Elimination.

$$x+y=4$$
$$y+z=7$$
$$3x-6y+21z=69$$

A. (1, 3, 0) B. (0, 3, 4) C. (1, 3, 4)

D. (3, 4, 0) E. (1, 1, 5)

6-12. Find the equation of the parabola through (0, 4), (1, 6), and (2, 9).

A. $y = x^2 + 3x - 4$

B. $y = x^2 + 3x$

C. $y = \frac{1}{2}x^2 + \frac{3}{2}x$

D. $y = x^2 + 3x + 4$

E. $y = \frac{1}{2}x^2 + \frac{3}{2}x + 4$

6-13. Solve (if possible) the system of three equations in three variables, using the method of Gauss-Jordan elimination.

$$2x + 3y + 2z = 3$$
$$4x + 5y + 3z = 6$$
$$4x + 7y + 2z = 9$$

A. $(1, 1, -1)$

B. $(2, 3, 2)$

C. $(-1, 1, 1)$

D. $\left(0, 0, \frac{3}{2}\right)$

E. No solution

6-14. Solve (if possible) the system of three equations in three variables, using the method of Gauss-Jordan elimination.

$$3x + 2y + 2z = 3$$
$$5x + 3y + 4z = 7$$
$$-2x - y - 2z = 5$$

A. $(1, 1, -1)$

B. $(3, 2, 2)$

C. $(-1, 1, 1)$

D. $\left(0, 0, \frac{3}{2}\right)$

E. No solution

6-15. Solve the system of nonlinear equations using the method of substitution.

$$x^2 + y^2 = 25$$
$$x = 3$$

A. (3, 0) B. (3, −4), (3, 4)

C. (3, 4) D. (3, −25), (3, 25)

E. No solution

6-16. Solve the system of nonlinear equations using the method of substitution.

$$3x^2 + 4y^2 = 7$$
$$x + y = 0$$

A. (−1, −1), (1, 1) B. (1, −1), (−1, 1)

C. (−1, −1), (−1, 1) D. (1, −1), (1, 1)

E. No solution

6-17. Solve the system of nonlinear equations using the method of elimination.

$$2x^2 + y^2 = 5$$
$$5x^2 - y^2 = 2$$

A. (−1, −1), (1, 1) B. $(1, \sqrt{3}), (-1, 3)$

C. $(1, \sqrt{3}), (1, -\sqrt{3}), (-1, \sqrt{3}) (-1, -\sqrt{3})$ D. (1, 3), (1, −3)

E. No solution

6-18. Determine which of the points (1, 1), (0, 12), (1, 13). and (1, 8) lie on the graph of the inequality $12x + y \le 13$.

A. (1, 1) and (1, 9)
B. (0, 12) and (1, 1)
C. (0, 12)
D. (1, 1), (0, 12) and (1, 9)
E. (1, 9) and (1, 13)

6-19. This is the graph of?

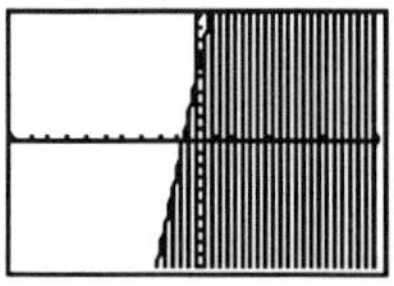

A. $-7x + y \le 5$
B. $7x - y \le 5$
C. $7x + y \le 5$
D. $-7x + y \ge 5$
E. $7x + y \ge 5$

6-20.

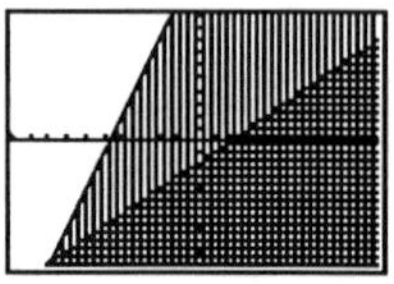

This graph represents the system of inequalities

A. $-3x + y \le 14$, $x - y \le -2$
B. $-3x + y \le 14$, $-x + y \ge 2$
C. $-3x + y \le 14$, $-x + y \le -2$
D. $3x - y \le 14$, $-x + y \le -2$
E. $3x - y \le 14$, $-x + y \ge 2$

6-21. Find the vertex of $2x + 3y \le 14$ and $x - y \le 2$.

A. (4, 2)
B. (−3, −1)
C. (1, 3)
D. (3, 1)
E. (−1, −3)

6-22. Find the vertices of

$3x+9y \leq 12$

$x+y \leq 4$

$x \geq 0$

$y \geq 0$

A. $\left(0, \frac{4}{3}\right)$, (4, 0), (0, 0) B. (4, 2), (0.25, 0), (0, 3), (0, 0)

C. (3, 1), (3, 0), (0, 0.25), (0, 0) D. (3, 1) only

E. (4, 2) and (0, 0) only

6-23. Maximize

$f = 5x+7y$

$4x+2y \leq 12$

$2x+4y \leq 12$

$x \geq 0,\ y \geq 0$

A. 6 B. 3 C. 9 D. 24 E. 20

6-24. At what point does the maximum occur?

$f = 6x+10y$

$4x+2y \leq 12$

$2x+4y \leq 12$

$x \geq 0,\ y \geq 0$

A. (0, 4) B. (0, 3) C. (6, 0)

D. (2, 2) E. (3, 0)

6-25. A person can grow and sell two different types of plants. Type I requires 4 ounces of fertilizer and 2 hours of time. Type II requires 2 ounces of fertilizer and 4 hours of time. There are 12 ounces of fertilizer and 12 hours of time available. If each plant of type I can be sold for \$5 and each plant of type II can be sold for \$7, what is the maximum amount of money the person can make?

A. \$9 B. \$24 C. \$20

D. \$6 E. \$3

Test form 6B   Date ________   Name ________________________
Course ______ Sec ______   Student Number ________________

Chapter 6

6-1. Solve the system of equations using the method of substitution.

$3x+4y=11$

$x+y=3$

A. (3, 4) B. (0, 0) C. (−1, 4)

D. (2, 1) E. (1, 2)

6-2. Solve the system of equations using the method of substitution.

$3x+4y=11$

$x+3y=7$

A. (3, 4) B. (0, 0) C. (−1, 4)

D. (2, 1) E. (1, 2)

6-3. Solve the system of equations using the method of elimination.

$5x+2y=12$

$2x+3y=7$

A. (0, 0) B. (5, 2) C. (−1, 4)

D. (2, 1) E. (1, 2)

6-4. Solve the system of equations using the method of elimination.

$5x+3y=11$

$4x-3y=-2$

A. (0, 0) B. (5, 3) C. (1, 2)

D. (−1, 4) E. (2, 1)

6-5. Solve the system of equations using the method of elimination. Give answer to four decimal places.

$5x+15y=40$

$3x+y=9$

A. (2.3750, 1.8750) B. (1.6429, 2.0714)

C. (3.0003, –0.0001) D. (0.4511, 8.5489)

E. (8.5489, 0.4511)

6-6. Solve the system of equations using the method of elimination. Give answer to four decimal places.

$3x+5y=8$

$-9x-15y=2$

A. (1, 1) B. (3, 5) C. (5, 3)

D. (–4, 4) E. No solution

6-7. A person has 14 coins with a value of $2.60. If the coins are all quarters and dimes, how many of each does the person have?

A. 8 quarters, 6 dimes B. 8 quarters, 8 dimes

C. 10 quarters, 8 dimes D. 10 quarters, 6 dimes

E. 0 quarters, 14 dimes

6-8. Determine the augmented matrix for the system

$4x+2y=7$

$6x+3y=9$

A. $\begin{bmatrix} 4 & 2 & 7 \\ 6 & 3 & 9 \end{bmatrix}$ B. $\begin{bmatrix} 4 & 6 \\ 2 & 3 \\ 7 & 9 \end{bmatrix}$ C. $\begin{bmatrix} 4 & 2 \\ 6 & 3 \\ 7 & 9 \end{bmatrix}$

D. $\begin{bmatrix} 4 & 2 \\ 6 & 3 \end{bmatrix}$ E. $\begin{bmatrix} 6 & 7 \\ 9 & 9 \end{bmatrix}$

6-9. Interpret the matrix as an augmented matrix of a system of equations. Write down the system.

$$\begin{bmatrix} 4 & 2 & 7 \\ 5 & 8 & 5 \end{bmatrix}$$

A. $\begin{aligned} 4x+5y&=7 \\ 2x+8y&=5 \end{aligned}$

B. $\begin{aligned} 4x+2y&=7 \\ 5x+8y&=5 \end{aligned}$

C. $\begin{aligned} 4x+2y+7z&=0 \\ 5x+8y+5z&=0 \end{aligned}$

D. $\begin{aligned} 2x+7y&=4 \\ 8x+5y&=5 \end{aligned}$

E. $\begin{aligned} 4x+7y&=2 \\ 5x+5y&=8 \end{aligned}$

6-10. Starting with $\begin{bmatrix} 1 & 0 & 2 & 5 \\ 3 & 4 & 5 & 7 \\ 8 & 5 & 7 & 4 \end{bmatrix}$, perform the transformation $-3R_1 + R_2$.

A. $\begin{bmatrix} 1 & 0 & 2 & 5 \\ 0 & 4 & -1 & -8 \\ 8 & 5 & 7 & 4 \end{bmatrix}$

B. $\begin{bmatrix} 1 & 0 & 2 & 5 \\ 6 & 4 & 11 & -8 \\ 8 & 5 & 7 & 4 \end{bmatrix}$

C. $\begin{bmatrix} 1 & 0 & 2 & 5 \\ 0 & 4 & 0 & -8 \\ 8 & 5 & 7 & 4 \end{bmatrix}$

D. $\begin{bmatrix} 3 & 4 & 5 & -8 \\ 8 & 5 & 7 & 4 \end{bmatrix}$

E. $\begin{bmatrix} 6 & 4 & 11 & -8 \\ 8 & 5 & 7 & 4 \end{bmatrix}$

6-11. Solve the system of equations using the method of Gauss-Jordan Elimination.

$$x + y = 4$$
$$y + z = 7$$
$$4x - 8y + 28z = 92$$

A. (1, 3, 0) B. (0, 3, 4) C. (1, 3, 4)

D. (3, 4, 0) E. (1, 1, 5)

6-12. Find the equation of the parabola through (0, 6), (1, 8), and (2, 11).

A. $y = x^2 + 3x - 6$

B. $y = x^2 + 3x + 6$

C. $y = \frac{1}{2}x^2 + \frac{3}{2}x + 6$

D. $y = \frac{1}{2}x^2 + \frac{3}{2}x$

E. $y = x^2 + 3x$

6-13. Solve (if possible) the system of three equations in three variables, using the method of Gauss-Jordan elimination.

$$3x + 4y + 2z = 5$$
$$5x + 3y + 2z = 6$$
$$6x + 2y + 3z = 5$$

A. (3, 4, 2)

B. (1, 1, −1)

C. (−1, 1, 1)

D. $\left(0, 0, \frac{5}{2}\right)$

E. No solution

6-14. Solve (if possible) the system of three equations in three variables, using the method of Gauss-Jordan elimination.

$$4x + 3y + 3z = 4$$
$$6x + 4y + 5z = 7$$
$$-2x - y - 2z = 3$$

A. (1, 1, −1)

B. (4, 3, 3)

C. (−1, 1, 1)

D. $\left(0, 0, \frac{4}{3}\right)$

E. No solution

6-15. Solve the system of nonlinear equations using the method of substitution.

$$x^2 + y^2 = 20$$
$$x = 2$$

A. (2, 0) B. (2, 4)

C. (2, –20), (2, 20) D. (2, –4), (2, 4)

E. No solution

6-16. Solve the system of nonlinear equations using the method of substitution.

$$5x^2 + 6y^2 = 11$$
$$x + y = 0$$

A. (–1, –1), (1, 1) B. (1, –1), (–1, 1)

C. (–1, –1), (–1, 1) D. (1, –1), (1, 1)

E. No solution

6-17. Solve the system of nonlinear equations using the method of elimination.

$$2x^2 + y^2 = 7$$
$$7x^2 - y^2 = 2$$

A. (–1, –1), (1, 1) B. $(1, \sqrt{5}), (-1, \sqrt{5})$

C. $(1, \sqrt{5}), (1, -\sqrt{5}), (-1, \sqrt{5})(-1, -\sqrt{5})$ D. (1, 5), (1, –5)

E. No solution

6-18. Determine which of the points (1, 1), (0, 13), (1, 14). and (1, 7) lie on the graph of the inequality $3x + y \leq 14$.

A. (1, 1) and (1, 9) B. (1, 14) and (1, 1)

C. (1, 1), (0, 13), and (1, 9) D. (0, 13)

E. (1, 9) and (1, 14)

6-19.

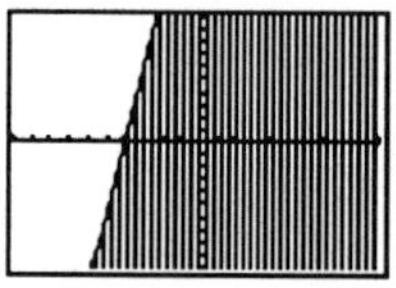

This is the graph of?

A. $-6x+y\le 2$ B. $6x-y\le 2$ C. $6x+y\le 2$

D. $-6x+y\ge 2$ E. $6x+y\ge 2$

6-20.

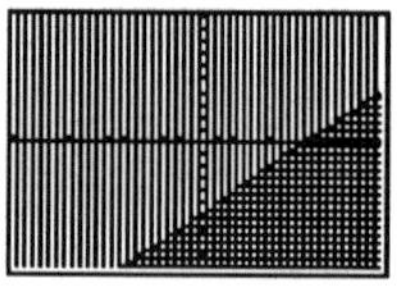

This graph represents the system of inequalities

A. $-2x+y\le 4$, $x-y\le -6$

B. $-2x+y\le 4$, $-x+y\ge 6$

C. $-2x+y\le 4$, $-x+y\le -6$

D. $2x-y\le 4$, $-x+y\le -6$

E. $2x-y\le 4$, $-x+y\ge 6$

6-21. Find the vertex of $2x+3y\le 14$ and $x-y\le 2$.

A. (4, 2) B. (−3, −1) C. (1, 3)

D. (3, 1) E. (−1, −3)

6-22. Find the vertices of

$3x+9y\le 12$

$x+y\le 4$

$x\ge 0$

$y\ge 0$

A. $\left(0, \frac{4}{3}\right)$, (4, 0), (0, 0)

B. (4, 2), (0.25, 0), (0, 3), (0, 0)

C. (3, 1), (0.25, 0), (0, 3), (0, 0)

D. (3, 1) only

E. (4, 2) and (0, 0) only

6-23. Maximize

$f = 5x+8y$

$4x+2y\le 12$

$2x+4y\le 12$

$x\ge 0,\ y\ge 0$

A. 26 B. 3 C. 24 D. 12 E. 25

6-24. At what point does the maximum occur?

$f = 4x+8y$

$4x+2y\le 12$

$2x+4y\le 12$

$y\le 2.5$

$x\ge 0,\ y\ge 0$

A. (0, 4)

B. (2, 2)

C. (6, 0)

D. (0, 3)

E. (3, 0)

6-25. A person can grow and sell two different types of plants. Type I requires 4 ounces of fertilizer and 2 hours of time. Type II requires 2 ounces of fertilizer and 4 hours of time. There are 12 ounces of fertilizer and 12 hours of time available. If each plant of type I can be sold for \$5 and each plant of type II can be sold for \$8, what is the maximum amount of money the person can make?

A. \$12 B. \$24 C. \$25 D. \$6 E. \$3

Test form 6C Date _________ Name _________________________
Course ______ Sec ______ Student Number ________________

Chapter 6

6-1. Solve the system of equations using the method of substitution.

$7x+2y=11$

$x+y=3$

A. (0, 0) B. (1, 2) C. (−1, 4)

D. (2, 1) E. (7, 2)

6-2. Solve the system of equations using the method of substitution.

$7x+2y=11$

$x+3y=7$

A. (0, 0) B. (1, 2) C. (−1, 4)

D. (2, 1) E. (7, 2)

6-3. Solve the system of equations using the method of elimination.

$6x+3y=15$

$2x+3y=7$

A. (2, 1) B. (6, 3) C. (−1, 4)

D. (0, 0) E. (1, 2)

6-4. Solve the system of equations using the method of elimination.

$6x+5y=16$

$4x-3y=-2$

A. (0, 0) B. (6, 5) C. (2, 1)

D. (−1, 4) E. (1, 2)

6-5. Solve the system of equations using the method of elimination. Give answer to four decimal places.

$2x+10y=16$

$3x+y=9$

A. (2.6429, 1.0714) B. (1.6429, 2.0714)

C. (3.0003, –0.0001) D. (0.4511, 8.5489)

E. (8.5489, 0.4511)

6-6. Solve the system of equations using the method of elimination. Give answer to four decimal places.

$3x+4y=13$

$-6x-8y=5$

A. (3, 4) B. (3, 1) C. (4, 3)

D. (7, –2) E. No solution

6-7. A person has 14 coins with a value of $2.90. If the coins are all quarters and dimes, how many of each does the person have?

A. 8 quarters, 4 dimes B. 8 quarters, 6 dimes

C. 10 quarters, 4 dimes D. 10 quarters, 6 dimes

E. 0 quarters, 14 dimes

6-8. Determine the augmented matrix for the system

$5x+6y=4$

$2x+3y=9$

A. $\begin{bmatrix} 5 & 6 & 4 \\ 2 & 3 & 9 \end{bmatrix}$ B. $\begin{bmatrix} 5 & 6 \\ 2 & 3 \\ 4 & 9 \end{bmatrix}$ C. $\begin{bmatrix} 5 & 2 \\ 6 & 3 \\ 4 & 9 \end{bmatrix}$

D. $\begin{bmatrix} 5 & 6 \\ 2 & 3 \end{bmatrix}$ E. $\begin{bmatrix} 11 & 4 \\ 5 & 9 \end{bmatrix}$

6-9. Interpret the matrix as an augmented matrix of a system of equations. Write down the system.

$$\begin{bmatrix} 3 & 4 & 7 \\ 2 & 8 & 6 \end{bmatrix}$$

A. $\begin{aligned} 3x+2y&=7 \\ 4x+8y&=6 \end{aligned}$

B. $\begin{aligned} 3x+4y&=7 \\ 2x+8y&=6 \end{aligned}$

C. $\begin{aligned} 3x+4y+7z&=0 \\ 2x+8y+6z&=0 \end{aligned}$

D. $\begin{aligned} 4x+7y&=3 \\ 8x+6y&=2 \end{aligned}$

E. $\begin{aligned} 3x+7y&=4 \\ 2x+6y&=8 \end{aligned}$

6-10. Starting with $\begin{bmatrix} 1 & 0 & 2 & 1 \\ 2 & 3 & 5 & 2 \\ 6 & 8 & 2 & 1 \end{bmatrix}$, perform the transformation $-2R_1 + R_2$.

A. $\begin{bmatrix} 1 & 0 & 2 & 1 \\ 4 & 6 & 10 & 4 \\ 6 & 8 & 2 & 1 \end{bmatrix}$

B. $\begin{bmatrix} 1 & 0 & 2 & 1 \\ 4 & 8 & 12 & 5 \\ 6 & 8 & 2 & 1 \end{bmatrix}$

C. $\begin{bmatrix} 1 & 0 & 2 & 1 \\ 0 & 3 & 1 & 0 \\ 6 & 8 & 2 & 1 \end{bmatrix}$

D. $\begin{bmatrix} 0 & 3 & 1 & 0 \\ 6 & 8 & 2 & 1 \end{bmatrix}$

E. $\begin{bmatrix} 4 & 8 & 12 & 5 \\ 6 & 8 & 2 & 1 \end{bmatrix}$

6-11. Solve the system of equations using the method of Gauss-Jordan Elimination.

$x + y = 4$

$y + z = 7$

$2x - 4y + 14z = 46$

A. (1, 3, 0) B. (0, 3, 4) C. (1, 3, 4)

D. (3, 4, 0) E. (1, 1, 5)

6-12. Find the equation of the parabola through (0, –4), (1, –2), and (2, 1).

A. $y = x^2 + 3x - 4$ B. $y = x^2 + 3x + 4$

C. $y = \frac{1}{2}x^2 + \frac{3}{2}x$ D. $y = \frac{1}{2}x^2 + \frac{3}{2}x - 4$

E. $y = x^2 + 3x$

6-13. Solve (if possible) the system of three equations in three variables, using the method of Gauss-Jordan elimination.

$$4x + 2y + 2z = 4$$
$$6x + 3y + z = 8$$
$$5x + 4y + 2z = 7$$

A. (4, 2, 2) B. (0, 0, 2) C. (–1, 1, 1)

D. (1, 1, –1) E. No solution

6-14. Solve (if possible) the system of three equations in three variables, using the method of Gauss-Jordan elimination.

$$5x + 3y + 4z = 7$$
$$-3x - 2y - 4z = 5$$
$$-2x - y = 3$$

A. (1, 1, –1) B. (5, 3, 4) C. (–1, 1, 1)

D. (0, –3, 0) E. No solution

6-15. Solve the system of nonlinear equations using the method of substitution.

$$x^2 + y^2 = 41$$
$$x = 4$$

A. (4, –5), (4, 5) B. (4, 0), (4, 5)

C. (4, 41) D. (4, –41), (4, 41)

E. No solution

6-16. Solve the system of nonlinear equations using the method of substitution.

$$2x^2 + 8y^2 = 10$$
$$x + y = 0$$

A. $(-1, -1), (1, 1)$ B. $(1, -1), (-1, 1)$

C. $(-1, -1), (-1, 1)$ D. $(1, -1), (1, 1)$

E. No solution

6-17. Solve the system of nonlinear equations using the method of elimination.

$$2x^2 + y^2 = 8$$
$$8x^2 - y^2 = 2$$

A. $(-1, -1), (1, 1)$ B. $(1, \sqrt{6}), (-1, \sqrt{6})$

C. $(1, \sqrt{6}), (1, -\sqrt{6}), (-1, \sqrt{6}) (-1, -\sqrt{6})$ D. $(1, 6), (1, -6)$

E. No solution

6-18. Determine which of the points (1, 1), (0, 6), (1, 3). and (1, 7) lie on the graph of the inequality $2x + y \leq 7$.

A. (1, 1) and (1, 3) B. (1, 3), (0, 6) and (1, 1)

C. (0, 6), and (1,7) D. (1, 1) and (0, 7)

E. (1, 1) and (0, 6)

6-19.

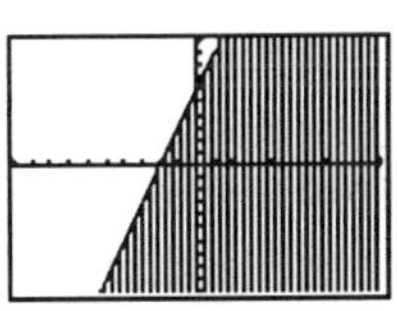

This is the graph of?

A. $-3x + y \leq 6$ B. $3x - y \leq 6$ C. $3x + y \leq 6$

D. $-3x + y \geq 6$ E. $3x + y \geq 6$

6-20.

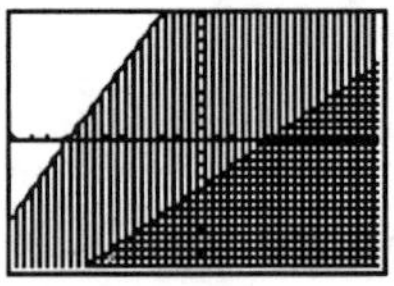

This graph represents the system of inequalities

A. $-2x + y \le 14$
$x - y \le -4$

B. $-2x + y \le 14$
$-x + y \ge 4$

C. $-2x + y \le 14$
$-x + y \le -4$

D. $2x - y \le 14$
$-x + y \le -4$

E. $2x - y \le 14$
$-x + y \ge 4$

6-21. Find the vertex of $4x + 2y \le 14$ and $x - y \le 2$.

A. (4, 2) B. (−3, −1) C. (1, 3)

D. (3, 1) E. (−1, −3)

6-22. Find the vertices of
$6x + 18y \le 24$
$x + y \le 4$
$x \ge 0$
$y \ge 0$

A. $\left(0, \dfrac{4}{3}\right)$, (4, 0), (0, 0) B. (4, 2), (0.25, 0), (0, 3), (0, 0)

C. (3, 1), (3, 0), (0, 0.25), (0, 0) D. (3, 1) only

E. (4, 2) and (0, 0) only

6-23. Maximize

$$f = 5x + 9y$$
$$4x + 2y \le 12$$
$$2x + 4y \le 12$$
$$x \ge 0,\ y \ge 0$$

A. 28 B. 3 C. 15 D. 27 E. 18

6-24. At what point does the maximum occur?

$$f = 5x + 9y$$
$$4x + 2y \le 12$$
$$2x + 4y \le 12$$
$$x \ge 0,\ y \ge 0$$

A. (0, 4) B. (2, 2) C. (6, 0)

D. (0, 3) E. (3, 0)

6-25. A person can grow and sell two different types of plants. Type I requires 4 ounces of fertilizer and 2 hours of time. Type II requires 2 ounces of fertilizer and 4 hours of time. There are 12 ounces of fertilizer and 12 hours of time available. If each plant of type I can be sold for \$5 and each plant of type II can be sold for \$9, what is the maximum amount of money the person can make?

A. \$15 B. \$27 C. \$18

D. \$28 E. \$3

Test form 6D Date ________ Name ________________________
Course ______ Sec ______ Student Number ______________

Chapter 6

6-1. Solve the system of equations using the method of substitution.

$6x+3y=12$

$x+y=3$

6-2. Solve the system of equations using the method of substitution.

$6x+3y=12$

$x+3y=7$

6-3. Solve the system of equations using the method of elimination.

$9x+3y=21$

$2x+3y=7$

6-4. Solve the system of equations using the method of elimination.

$4x+5y=13$

$4x-3y=5$

6-5. Solve the system of equations using the method of elimination. Give answer to four decimal places.

$3x+15y=24$

$3x+y=9$

6-6. Solve the system of equations using the method of elimination. Give answer to four decimal places.

$4x+9y=7$

$-16x-36y=5$

6-7. A person has 7 coins with a value of $1.30. If the coins are all quarters and dimes, how many of each does the person have?

6-8. Determine the augmented matrix for the system

$7x+2y=5$

$4x+3y=7$

6-9. Interpret the matrix as an augmented matrix of a system of equations. Write down the system.

$$\begin{bmatrix} 5 & 4 & 6 \\ 2 & 3 & 7 \end{bmatrix}$$

6-10. Starting with $\begin{bmatrix} 1 & 0 & 4 & 3 \\ 2 & 1 & 0 & 3 \\ 1 & 7 & 2 & 6 \end{bmatrix}$, perform the transformation $-2R_1 + R_2$.

6-11. Solve the system of equations using the method of Gauss-Jordan Elimination.

$$x + y = 4$$
$$y + z = 7$$
$$5x - 10y + 35z = 115$$

6-12. Find the equation of the parabola through (0, 8), (1, 10), and (2, 13).

6-13. Solve (if possible) the system of three equations in three variables, using the method of Gauss-Jordan elimination.

$$5x + 7y + 2z = 10$$
$$3x + 5y + z = 7$$
$$7x + 6y + 2z = 11$$

6-14. Solve (if possible) the system of three equations in three variables, using the method of Gauss-Jordan elimination.

$$2x + 5y + 7z = 1$$
$$4x + 2y + 3z = 2$$
$$-2x + 3y + 4z = 1$$

6-15. Solve the system of nonlinear equations using the method of substitution.

$$x^2 + y^2 = 29$$
$$x = 2$$

6-16. Solve the system of nonlinear equations using the method of substitution.

$$5x^2 + 3y^2 = 8$$
$$x + y = 0$$

6-17. Solve the system of nonlinear equations using the method of elimination.

$$2x^2 + y^2 = 10$$
$$10x^2 - y^2 = 2$$

6-18. Determine which of the points (1, 1), (0, 9), (1, 4). and (1, 10) lie on the graph of the inequality $2x + y \leq 10$.

6-19.

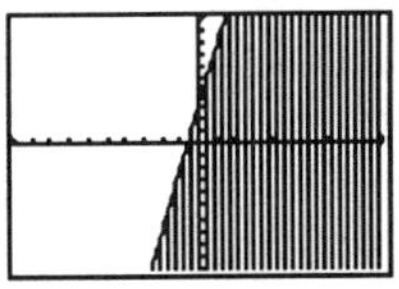

This is the graph of?

6-20.

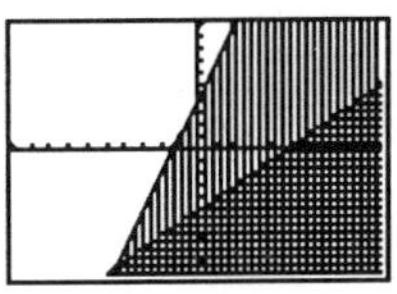

This graph represents the system of inequalities

6-21. Find the vertex of $7x + 5y \leq 12$ and $x + y \leq 2$.

6-22. Find the vertices of

$$x + 3y \leq 4$$
$$x + y \leq 4$$
$$x \geq 0$$
$$y \geq 0$$

6-23. Maximize

$$f = 7x + 11y$$
$$4x + 2y \leq 12$$
$$2x + 4y \leq 12$$
$$x \geq 0,\ y \geq 0$$

6-24. At what point does the maximum occur?

$$f = 7x + 11y$$
$$4x + 2y \leq 12$$
$$2x + 4y \leq 12$$
$$x \geq 0,\ y \geq 0$$

6-25. A person can grow and sell two different types of plants. Type I requires 4 ounces of fertilizer and 2 hours of time. Type II requires 2 ounces of fertilizer and 4 hours of time. There are 12 ounces of fertilizer and 12 hours of time available. If each plant of type I can be sold for \$7 and each plant of type II can be sold for \$11, what is the maximum amount of money the person can make?

Test form 6E Date __________ Name ____________________________
Course ______ Sec ______ Student Number __________________

Chapter 6

6-1. Solve the system of equations using the method of substitution.

$5x+4y=13$

$x+y=3$

6-2. Solve the system of equations using the method of substitution.

$5x+4y=13$

$x+3y=7$

6-3. Solve the system of equations using the method of elimination.

$5x+6y=16$

$2x+3y=7$

6-4. Solve the system of equations using the method of elimination.

$7x+2y=11$

$4x-3y=-2$

6-5. Solve the system of equations using the method of elimination. Give answer to four decimal places.

$8x+40y=64$

$3x+y=9$

6-6. Solve the system of equations using the method of elimination. Give answer to four decimal places.

$7x+3y=6$

$-21x-9y=7$

6-7. A person has 9 coins with a value of \$1.80. If the coins are all quarters and dimes, how many of each does the person have?

6-8. Determine the augmented matrix for the system

$9x+2y=1$

$3x+4y=7$

6-9. Interpret the matrix as an augmented matrix of a system of equations. Write down the system.

$$\begin{bmatrix} 3 & 9 & 2 \\ 8 & 7 & 4 \end{bmatrix}$$

6-10. Starting with $\begin{bmatrix} 1 & 0 & 1 & 1 \\ 4 & 1 & 1 & 2 \\ 6 & 8 & 1 & 7 \end{bmatrix}$, perform the transformation $-4R_1 + R_2$.

6-11. Solve the system of equations using the method of Gauss-Jordan Elimination.

$$x + y = 4$$
$$y + z = 7$$
$$7x - 14y + 49z = 161$$

6-12. Find the equation of the parabola through (0, 10), (1, 12), and (2, 15).

6-13. Solve (if possible) the system of three equations in three variables, using the method of Gauss-Jordan elimination.

$$6x + 5y + 2z = 9$$
$$4x + 3y + 5z = 2$$
$$6x + 2y + 3z = 5$$

6-14. Solve (if possible) the system of three equations in three variables, using the method of Gauss-Jordan elimination.

$$4x + 3y + 6z = 1$$
$$3x + 2y + 5z = 8$$
$$-x - y - z = 2$$

6-15. Solve the system of nonlinear equations using the method of substitution.

$$x^2 + y^2 = 53$$
$$x = 9$$

6-16. Solve the system of nonlinear equations using the method of substitution.

$$4x^2 + 2y^2 = 6$$
$$x + y = 0$$

6-17. Solve the system of nonlinear equations using the method of elimination.

$$2x^2 + y^2 = 12$$
$$12x^2 - y^2 = 2$$

6-18. Determine which of the points (1, 1), (1, 5), (0, 11). and (1, 12) lie on the graph of the inequality $2x + y \le 11$.

6-19.

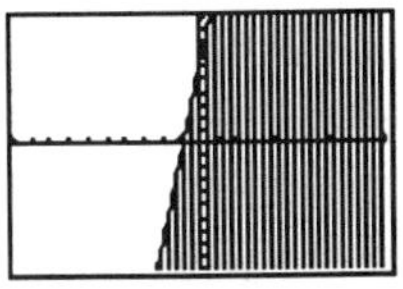

This is the graph of?

6-20.

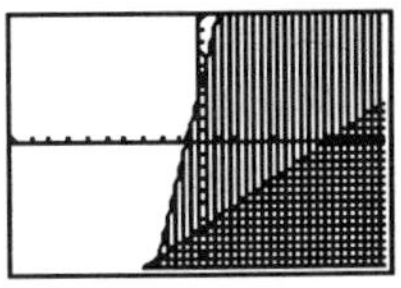

This graph represents the system of inequalities

6-21. Find the vertex of $8x + 6y \le 14$ and $x + y \le 2$.

6-22. Find the vertices of

$$2x + 6y \le 8$$
$$x + y \le 4$$
$$x \ge 0$$
$$y \ge 0$$

6-23. Maximize

$$f = 9x + 13y$$
$$4x + 2y \le 12$$
$$2x + 4y \le 12$$
$$x \ge 0,\ y \ge 0$$

6-24. At what point does the maximum occur?

$$f = 9x + 13y$$
$$4x + 2y \le 12$$
$$2x + 4y \le 12$$
$$x \ge 0,\ y \ge 0$$

6-25. A person can grow and sell two different types of plants. Type I requires 4 ounces of fertilizer and 2 hours of time. Type II requires 2 ounces of fertilizer and 4 hours of time. There are 12 ounces of fertilizer and 12 hours of time available. If each plant of type I can be sold for $9 and each plant of type II can be sold for $13, what is the maximum amount of money the person can make?

Test form 6F Date ________ Name ________________________
Course ______ Sec ______ Student Number ______________

Chapter 6

6-1. Solve the system of equations using the method of substitution.

$$9x + 3y = 15$$
$$x + y = 3$$

6-2. Solve the system of equations using the method of substitution.

$$9x + 3y = 15$$
$$x + 3y = 7$$

6-3. Solve the system of equations using the method of elimination.

$$5x + 8y = 18$$
$$2x + 3y = 7$$

6-4. Solve the system of equations using the method of elimination.

$$6x + 3y = 12$$
$$4x - 3y = -2$$

6-5. Solve the system of equations using the method of elimination. Give answer to four decimal places.

$$6x + 30y = 48$$
$$3x + y = 9$$

6-6. Solve the system of equations using the method of elimination. Give answer to four decimal places.

$$8x + 2y = 5$$
$$-4x - y = 1$$

6-7. A person has 10 coins with a value of $1.90. If the coins are all quarters and dimes, how many of each does the person have?

6-8. Determine the augmented matrix for the system

$$2x + 7y = 3$$
$$4x + 5y = 1$$

6-9. Interpret the matrix as an augmented matrix of a system of equations. Write down the system.

$$\begin{bmatrix} 9 & 7 & 8 \\ 3 & 6 & 5 \end{bmatrix}$$

6-10. Starting with $\begin{bmatrix} 1 & 0 & 2 & 1 \\ 5 & 1 & 0 & 1 \\ 3 & 7 & 2 & 6 \end{bmatrix}$, perform the transformation $-5R_1 + R_2$.

6-11. Solve the system of equations using the method of Gauss-Jordan Elimination.

$$x + y = 4$$
$$y + z = 7$$
$$10x - 20y + 70z = 230$$

6-12. Find the equation of the parabola through (0, 2), (1, 4), and (2, 7).

6-13. Solve (if possible) the system of three equations in three variables, using the method of Gauss-Jordan elimination.

$$4x + 6y + 2z = 8$$
$$3x + 5y + 3z = 5$$
$$7x + 2y + 8z = 1$$

6-14. Solve (if possible) the system of three equations in three variables, using the method of Gauss-Jordan elimination.

$$2x + 4y + 5z = 7$$
$$3x + 5y + 6z = 9$$
$$x + y + z = 5$$

6-15. Solve the system of nonlinear equations using the method of substitution.

$$x^2 + y^2 = 26$$
$$x = 1$$

6-16. Solve the system of nonlinear equations using the method of substitution.

$$7x^2 + 2y^2 = 9$$
$$x + y = 0$$

6-17. Solve the system of nonlinear equations using the method of elimination.

$$2x^2 + y^2 = 3$$
$$3x^2 - y^2 = 2$$

6-18. Determine which of the points (1, 1), (0, 15), (1, 7). and (1, 16) lie on the graph of the inequality $2x + y \leq 16$.

6-19.

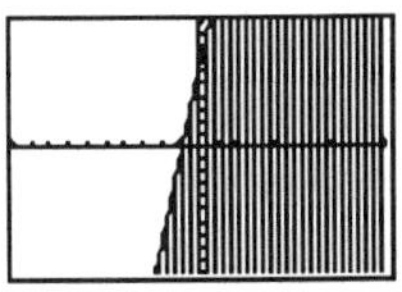

This is the graph of?

6-20.

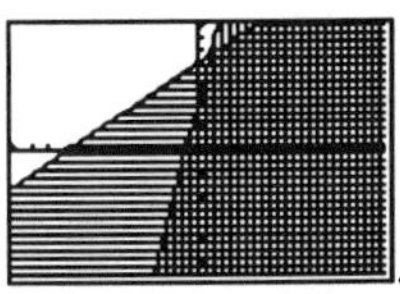

This graph represents the system of inequalities

6-21. Find the vertex of $4x + 2y \leq 6$ and $x + y \leq 2$.

6-22. Find the vertices of
$7x + 21y \leq 28$
$x + y \leq 4$
$x \geq 0$
$y \geq 0$

6-23. Maximize

$f = 6x + 10y$
$4x + 2y \leq 12$
$2x + 4y \leq 12$
$x \geq 0,\ y \geq 0$

6-24. At what point does the maximum occur?

$$f = 6x + 10y$$
$$4x + 2y \leq 12$$
$$2x + 4y \leq 12$$
$$x \geq 0,\ y \geq 0$$

6-25. A person can grow and sell two different types of plants. Type I requires 4 ounces of fertilizer and 2 hours of time. Type II requires 2 ounces of fertilizer and 4 hours of time. There are 12 ounces of fertilizer and 12 hours of time available. If each plant of type I can be sold for \$6 and each plant of type II can be sold for \$10, what is the maximum amount of money the person can make?

Test form 7A Date _________ Name ____________________________
Course ______ Sec ______ Student Number ________________

Chapter 7

7-1. Give the size of the matrix $\begin{bmatrix} 5 & 2 & 1 \\ 6 & 8 & 4 \end{bmatrix}$.

A. 4 B. 9 C. 6

D. 3×2 E. 2×3

7-2. For A = $\begin{bmatrix} -6 & -8 & -9 \\ 5 & -2 & -5 \\ 5 & 2 & 5 \end{bmatrix}$, find $a_{3,2}$.

A. −1 B. −2 C. −3 D. 2 E. 5

7-3. Solve the matrix equation $\begin{bmatrix} x & x+y \\ y+z & z \end{bmatrix} = \begin{bmatrix} 6 & 8 \\ 9 & 7 \end{bmatrix}$.

A. (6, 2, 7) B. (6, –3, 7) C. (6, –2, 7)

D. (6, 0, 7) E. (6, 3, 7)

7-4. A = $\begin{bmatrix} 5 & 6 \\ 3 & 2 \end{bmatrix}$, B = $\begin{bmatrix} 4 & 0 \\ 1 & 2 \end{bmatrix}$ Perform the operations A + 2B, if possible.

A. $\begin{bmatrix} 13 & 0 \\ 5 & 6 \end{bmatrix}$ B. $\begin{bmatrix} 13 & 0 \\ 5 & 2 \end{bmatrix}$ C. $\begin{bmatrix} 13 & 6 \\ 5 & 2 \end{bmatrix}$

D. $\begin{bmatrix} 13 & 6 \\ 5 & 6 \end{bmatrix}$ E. Not possible

7-5. $A = \begin{bmatrix} 1 & 2 & 0 & 1 \\ 3 & 5 & 9 & 2 \\ 1 & 1 & 0 & 2 \\ 0 & 1 & 4 & 1 \end{bmatrix}$ If $a_{i,j} = 9$, what are the possible values of j?

A. 3 B. 2 C. 2, 3

D. 1, 2, 3 E. 2, 3, 4

7-6. Perform the multiplication $\begin{bmatrix} 5 & 1 \\ 2 & 4 \end{bmatrix}\begin{bmatrix} 3 & 0 \\ 1 & 4 \end{bmatrix}$, if possible.

A. $\begin{bmatrix} 16 & 4 \\ 10 & 16 \end{bmatrix}$ B. $\begin{bmatrix} 16 & 4 \\ 26 & 16 \end{bmatrix}$ C. $\begin{bmatrix} 16 & 9 \\ 26 & 4 \end{bmatrix}$

D. $\begin{bmatrix} 16 & 4 \\ 10 & 4 \end{bmatrix}$ E. Not possible

7-7. If A is a matrix of size 4×3 and B is a matrix of size 3×4, what size is AB, if it exists?

A. 3×3 B. 3×4 C. 4×4

D. 4×3 E. Does not exist.

7-8. A is a matrix of size 7×3, B of size 3×7, and C of size 3×3. What is the size of 2AB – 3C, if it exists.

A. 7×7 B. 7×3 C. 3×3

D. 3×7 E. Does not exist.

7-9. If $A = \begin{bmatrix} 1 & 0 & 1 \\ 3 & 1 & 7 \\ 2 & 1 & 4 \end{bmatrix}$, what is I such that AI = A?

A. $\begin{bmatrix} 0 & 1 \\ 1 & 0 \end{bmatrix}$

B. $\begin{bmatrix} 1 & 0 \\ 0 & 1 \end{bmatrix}$

C. $\begin{bmatrix} 1 & 0 & 0 \\ 0 & 1 & 0 \\ 0 & 0 & 1 \end{bmatrix}$

D. $\begin{bmatrix} 1 & 0 & 1 \\ 0 & 1 & 0 \\ 1 & 0 & 1 \end{bmatrix}$

E. $\begin{bmatrix} 0 & 0 & 1 \\ 0 & 1 & 0 \\ 1 & 0 & 0 \end{bmatrix}$

7-10. Determine the inverse of $\begin{bmatrix} 1 & 2 \\ 0 & 1 \end{bmatrix}$, if it exists.

A. $\begin{bmatrix} \frac{1}{2} & -1 \\ 0 & \frac{1}{2} \end{bmatrix}$

B. $\begin{bmatrix} 1 & -2 \\ 0 & 1 \end{bmatrix}$

C. $\begin{bmatrix} 2 & -4 \\ 0 & 2 \end{bmatrix}$

D. $\begin{bmatrix} \frac{1}{2} & 1 \\ 0 & \frac{1}{2} \end{bmatrix}$

E. Does not exit.

7-11. Determine the inverse of $\begin{bmatrix} 6 & 1 & 0 \\ 2 & 3 & 4 \\ 1 & 1 & 0 \end{bmatrix}$, if it exists.

A. $\begin{bmatrix} .2 & -.2 & 0 \\ -.2 & 1.2 & 0 \\ .05 & .25 & -.8 \end{bmatrix}$

B. $\begin{bmatrix} .2 & 1 & -.2 \\ -.2 & 1 & 1.2 \\ .05 & .25 & -.8 \end{bmatrix}$

C. $\begin{bmatrix} -.2 & 0 & .2 \\ .2 & 0 & 1.2 \\ .05 & .25 & -.8 \end{bmatrix}$

D. $\begin{bmatrix} .2 & 0 & -.2 \\ -.2 & 0 & 1.2 \\ .05 & .25 & -.8 \end{bmatrix}$

E. Does not exist.

7-12. Solve the system of equations using the matrix inverse method.

$$2x+6y=18$$
$$3x+12y=27$$

A. (18, 27) B. (3, 12) C. (9, 0)

D. (2, 6) E. Does not exist.

7-13. Solve the system of equations using the matrix inverse method.

$$5x+2y+z=20$$
$$4x-y+3z=13$$
$$5x+y-6z=11$$

A. (3, 2, 1) B. (1, 0, 3) C. (2, 1, 0)

D. (2, −1, 2) E. Does not exist.

7-14. Consider the economy defined by the input-output matrix A and open sector D. Determine the output levels required of the industries to meet the demands of the industries and the open sector.

$$A=\begin{bmatrix} .2 & .5 \\ .4 & .1 \end{bmatrix},\ D=\begin{bmatrix} 50 \\ 70 \end{bmatrix}$$

A. $\begin{bmatrix} 5 \\ -43 \end{bmatrix}$ B. $\begin{bmatrix} 154 \\ 146 \end{bmatrix}$ C. $\begin{bmatrix} 50 \\ 70 \end{bmatrix}$

D. $\begin{bmatrix} 19 \\ 69 \end{bmatrix}$ E. $\begin{bmatrix} 45 \\ -15 \end{bmatrix}$

7-15. Compute the determinant $\begin{vmatrix} 2 & 3 \\ 4 & 5 \end{vmatrix}$

A. −2 B. 12 C. 11 D. −11 E. 22

7-16. Compute the determinant $\begin{vmatrix} 7 & -7 & 7 \\ 1 & 2 & 3 \\ 4 & 5 & 6 \end{vmatrix}$.

A. −12 B. 0 C. 12 D. −84 E. 84

7-17. Compute the minor $a_{3,1}$ of $A = \begin{vmatrix} 2 & 4 & 2 \\ 2 & 4 & 2 \\ 2 & 2 & 4 \end{vmatrix}$.

A. 2 B. 4 C. −4 D. 0 E. −2

7-18. Solve the system using Cramer's rule if it is applicable. If Cramer's rule cannot be used say so.

$$5x + 8y = 47$$
$$7x - 2y = 13$$

A. (5, 11) B. (14, 0.5) C. (3, 4)

D. (5, 8) E. Cramer's rule cannot be used.

7-19. Solve the system using Cramer's rule if it is applicable. If Cramer's rule cannot be used say so.

$$4x - 2y = 6$$
$$-8x + 4y = 7$$

A. (1, −1) B. (3, 4) C. (1, 1)

D. (−1, 1) E. Cramer's rule cannot be used.

7-20. Solve the system using Cramer's rule if it is applicable. If Cramer's rule cannot be used say so.

$$3x + 4y + z = 1$$
$$x + y - 2z = 5$$
$$7x + 9y = 7$$

A. (1, 2, −1) B. (1, 2, 0) C. (2, 3, 0)

D. (4, 1, 0) E. Cramer's rule cannot be used.

7-21. Compute the determinant of the matrix $\begin{bmatrix} 2 & 5 \\ 3 & 2 \end{bmatrix}$.

A. 11 B. −11 C. 19 D. −19 E. 0

7-22. Compute the determinant of the matrix $\begin{bmatrix} 5 & 5 & 5 \\ 2 & 3 & 4 \\ 5 & 6 & 7 \end{bmatrix}$.

A. 0 B. 5 C. −5 D. −1 E. 1

7-23. $A = \begin{bmatrix} 5 & 4 \\ 2 & 8 \end{bmatrix}$. Determine the transpose of A.

A. $\begin{bmatrix} 5 & 2 \\ 4 & 8 \end{bmatrix}$ B. $\begin{bmatrix} 5 & 4 \\ 2 & 8 \end{bmatrix}$ C. $\begin{bmatrix} 8 & 2 \\ 4 & 5 \end{bmatrix}$

D. $\begin{bmatrix} 8 & 4 \\ 2 & 5 \end{bmatrix}$ E. $\begin{bmatrix} 5 & 8 \\ 4 & 2 \end{bmatrix}$

7-24. The matrix $A = \begin{bmatrix} 5 & 2 & 3 \\ 2 & 4 & * \\ 3 & 1 & 2 \end{bmatrix}$ is to be symmetric. Determine the element indicated *.

A. 3 B. 1 C. 4 D. 2 E. 5

7-25. Compute $g_{3,1}$ of the matrix $G = AA'$ where $A = \begin{bmatrix} 1 & 1 & 0 \\ 0 & 0 & 1 \\ 0 & 1 & 1 \end{bmatrix}$.

A. 2 B. 1 C. 0 D. 9 E. 3

Test form 7B　　Date ________　　Name ________________________
Course ______ Sec ______　　Student Number ________________

Chapter 7

7-1. Give the size of the matrix $\begin{bmatrix} 3 & 2 & 7 \\ 1 & 5 & 4 \end{bmatrix}$.

A. 4　　B. 9　　C. 6

D. 3×2　　E. 2×3

7-2. For $A = \begin{bmatrix} -6 & -8 & -9 \\ 7 & -2 & -7 \\ 7 & 2 & 7 \end{bmatrix}$, find $a_{3,2}$.

A. –1　　B. –2　　C. –3　　D. 7　　E. 2

7-3. Solve the matrix equation $\begin{bmatrix} x & x+y \\ y+z & z \end{bmatrix} = \begin{bmatrix} 1 & 3 \\ 7 & 5 \end{bmatrix}$.

A. (1, 2, 5)　　B. (1, –3, 5)　　C. (1, –2, 5)

D. (1, 0, 5)　　E. (1, 3, 5)

7-4. $A = \begin{bmatrix} 1 & 4 \\ 7 & 9 \end{bmatrix}$, $B = \begin{bmatrix} 4 & 0 \\ 1 & 2 \end{bmatrix}$ Perform the operations A + 2B, if possible.

A. $\begin{bmatrix} 9 & 4 \\ 11 & 11 \end{bmatrix}$　　B. $\begin{bmatrix} 9 & 0 \\ 11 & 13 \end{bmatrix}$　　C. $\begin{bmatrix} 9 & 0 \\ 11 & 11 \end{bmatrix}$

D. $\begin{bmatrix} 9 & 4 \\ 9 & 13 \end{bmatrix}$　　E. Not possible

7-5. $A = \begin{bmatrix} 1 & 2 & 0 & 1 \\ 3 & 5 & 6 & 2 \\ 1 & 1 & 0 & 2 \\ 0 & 1 & 4 & 1 \end{bmatrix}$ If $a_{i,j} = 6$, what are the possible values of j?

A. 2 B. 3 C. 2, 3

D. 1, 2, 3 E. 2, 3, 4

7-6. Perform the multiplication $\begin{bmatrix} 6 & 1 \\ 2 & 4 \end{bmatrix}\begin{bmatrix} 3 & 0 \\ 1 & 4 \end{bmatrix}$, if possible.

A. $\begin{bmatrix} 19 & 4 \\ 35 & 16 \end{bmatrix}$ B. $\begin{bmatrix} 19 & 4 \\ 10 & 16 \end{bmatrix}$ C. $\begin{bmatrix} 19 & 7 \\ 10 & 4 \end{bmatrix}$

D. $\begin{bmatrix} 19 & 7 \\ 35 & 4 \end{bmatrix}$ E. Not possible

7-7. If A is a matrix of size 5×2 and B is a matrix of size 2×5, what size is AB, if it exists?

A. 2×2 B. 2×5 C. 5×2

D. 5×5 E. Does not exist.

7-8. A is a matrix of size 2×3, B of size 3×2, and C of size 3×3. What is the size of $2AB - 3C$, if it exists.

A. 2×2 B. 2×3 C. 3×3

D. 3×3 E. Does not exist.

7-9. If $A = \begin{bmatrix} 7 & 1 & 3 \\ 4 & 1 & 1 \\ 2 & 0 & 1 \end{bmatrix}$, what is I such that AI = A?

A. $\begin{bmatrix} 0 & 1 \\ 1 & 0 \end{bmatrix}$ B. $\begin{bmatrix} 1 & 0 \\ 0 & 1 \end{bmatrix}$ C. $\begin{bmatrix} 1 & 0 & 0 \\ 0 & 1 & 0 \\ 0 & 0 & 1 \end{bmatrix}$

D. $\begin{bmatrix} 1 & 0 & 1 \\ 0 & 1 & 0 \\ 1 & 0 & 1 \end{bmatrix}$ E. $\begin{bmatrix} 0 & 0 & 1 \\ 0 & 1 & 0 \\ 1 & 0 & 0 \end{bmatrix}$

7-10. Determine the inverse of $\begin{bmatrix} 1 & 3 \\ 0 & 1 \end{bmatrix}$, if it exists.

A. $\begin{bmatrix} 1 & -3 \\ 0 & 1 \end{bmatrix}$ B. $\begin{bmatrix} \frac{1}{3} & -\frac{2}{3} \\ 0 & \frac{1}{3} \end{bmatrix}$ C. $\begin{bmatrix} 3 & -6 \\ 0 & 3 \end{bmatrix}$

D. $\begin{bmatrix} \frac{1}{3} & \frac{2}{3} \\ 0 & \frac{1}{3} \end{bmatrix}$ E. Does not exit.

7-11. Determine the inverse of $\begin{bmatrix} 9 & 1 & 4 \\ 0 & 2 & 1 \\ 1 & 1 & 0 \end{bmatrix}$, if it exists.

A. $\begin{bmatrix} .063 & -.25 & .438 \\ -.063 & .25 & .563 \\ .125 & .5 & -1.125 \end{bmatrix}$ B. $\begin{bmatrix} -.067 & -.267 & -.467 \\ -.067 & -.267 & -.533 \\ -.133 & -.467 & 1.067 \end{bmatrix}$

C. $\begin{bmatrix} .087 & -.287 & .493 \\ -.087 & .287 & .614 \\ .144 & .472 & .134 \end{bmatrix}$ D. $\begin{bmatrix} .111 & .124 & .891 \\ .317 & .002 & .002 \\ .817 & .031 & .911 \end{bmatrix}$

E. Does not exist.

7-12. Solve the system of equations using the matrix inverse method.

$$4x + 12y = 36$$
$$3x + 12y = 33$$

A. (3, 2) B. (4, 12) C. (36, 27)

D. (3, 12) E. Does not exist.

7-13. Solve the system of equations using the matrix inverse method.

$$7x + 2y + z = 26$$
$$4x - y + 3z = 13$$
$$5x + y - 6z = 11$$

A. (3, 2, 1) B. (1, 0, 3) C. (2, 1, 0)

D. (2, −1, 2) E. Does not exist.

7-14. Consider the economy defined by the input-output matrix A and open sector D. Determine the output levels required of the industries to meet the demands of the industries and the open sector.

$$A = \begin{bmatrix} .2 & .5 \\ .4 & .1 \end{bmatrix}, D = \begin{bmatrix} 381 \\ 410 \end{bmatrix}$$

A. $\begin{bmatrix} 10 \\ -86 \end{bmatrix}$ B. $\begin{bmatrix} 100 \\ 140 \end{bmatrix}$ C. $\begin{bmatrix} 1054 \\ 924 \end{bmatrix}$

D. $\begin{bmatrix} 90 \\ -30 \end{bmatrix}$ E. $\begin{bmatrix} 38 \\ 138 \end{bmatrix}$

7-15. Compute the determinant $\begin{vmatrix} 4 & 2 \\ 5 & 6 \end{vmatrix}$

A. −14 B. 14 C. −13 D. 13 E. 34

7-16. Compute the determinant $\begin{vmatrix} 4 & -4 & 4 \\ 1 & 2 & 3 \\ 4 & 5 & 6 \end{vmatrix}$.

A. −12 B. 0 C. 12 D. −48 E. 48

7-17. Compute the minor $a_{3,1}$ of A = $\begin{vmatrix} 3 & 6 & 3 \\ 3 & 6 & 3 \\ 3 & 3 & 6 \end{vmatrix}$.

A. 3 B. 6 C. −6 D. −3 E. 0

7-18. Solve the system using Cramer's rule if it is applicable. If Cramer's rule cannot be used say so.

$$2x + 4y = 22$$
$$7x - 2y = 13$$

A. (5, 11) B. (14, 0.5) C. (3, 4)

D. (2, 4) E. Cramer's rule cannot be used.

7-19. Solve the system using Cramer's rule if it is applicable. If Cramer's rule cannot be used say so.

$$9x - 3y = 5$$
$$-3x + y = 6$$

A. (1, −1) B. (3, 4) C. (1, 1)

D. (−1, 1) E. Cramer's rule cannot be used.

7-20. Solve the system using Cramer's rule if it is applicable. If Cramer's rule cannot be used say so.

$$2x + 3y + z = 5$$
$$x + y - 2z = 5$$
$$5x + 7y = 15$$

A. (1, 2, −1) B. (1, 2, 0) C. (2, 3, 0)

D. (4, 1, 0) E. Cramer's rule cannot be used.

7-21. Compute the determinant of the matrix $\begin{bmatrix} 3 & 5 \\ 2 & 6 \end{bmatrix}$.

A. 8 B. –8 C. 28 D. –28 E. 0

7-22. Compute the determinant of the matrix $\begin{bmatrix} 6 & 6 & 6 \\ 2 & 3 & 4 \\ 5 & 6 & 7 \end{bmatrix}$.

A. 0 B. 6 C. –6 D. –1 E. 1

7-23. $A = \begin{bmatrix} 7 & 1 \\ 6 & 3 \end{bmatrix}$. Determine the transpose of A.

A. $\begin{bmatrix} 6 & 3 \\ 7 & 1 \end{bmatrix}$ B. $\begin{bmatrix} 7 & 6 \\ 1 & 3 \end{bmatrix}$ C. $\begin{bmatrix} 3 & 1 \\ 6 & 7 \end{bmatrix}$

D. $\begin{bmatrix} 3 & 6 \\ 1 & 7 \end{bmatrix}$ E. $\begin{bmatrix} 7 & 1 \\ 6 & 3 \end{bmatrix}$

7-24. The matrix $A = \begin{bmatrix} 5 & 2 & 4 \\ 2 & 6 & * \\ 4 & 2 & 1 \end{bmatrix}$ is to be symmetric. Determine the element indicated *.

A. 4 B. 2 C. 1 D. 6 E. 5

7-25. Compute $g_{3,2}$ of the matrix $G = AA'$ where $A = \begin{bmatrix} 1 & 1 & 0 \\ 0 & 0 & 1 \\ 0 & 1 & 1 \end{bmatrix}$.

A. 2 B. 1 C. 0 D. 9 E. 3

Test form 7C Date _________ Name _________________________
Course ______ Sec ______ Student Number ________________

Chapter 7

7-1. Give the size of the matrix $\begin{bmatrix} 8 & 7 & 3 \\ 7 & 4 & 1 \end{bmatrix}$.

A. 4 B. 9 C. 6

D. 3×2 E. 2×3

7-2. For $A = \begin{bmatrix} -6 & -8 & -9 \\ 4 & -1 & -4 \\ 4 & 1 & 4 \end{bmatrix}$, find $a_{3,2}$.

A. 1 B. −2 C. −3 D. −1 E. 4

7-3. Solve the matrix equation $\begin{bmatrix} x & x+y \\ y+z & z \end{bmatrix} = \begin{bmatrix} 3 & 5 \\ 6 & 4 \end{bmatrix}$.

A. (3, 2, 4) B. (3, −3, 4) C. (3, −2, 4)

D. (3, 0, 4) E. (3, 3, 4)

7-4. $A = \begin{bmatrix} 1 & 6 \\ 5 & 9 \end{bmatrix}$, $B = \begin{bmatrix} 4 & 0 \\ 1 & 2 \end{bmatrix}$ Perform the operations A + 2B, if possible.

A. $\begin{bmatrix} 9 & 6 \\ 7 & 13 \end{bmatrix}$ B. $\begin{bmatrix} 9 & 0 \\ 7 & 9 \end{bmatrix}$ C. $\begin{bmatrix} 9 & 6 \\ 7 & 9 \end{bmatrix}$

D. $\begin{bmatrix} 9 & 0 \\ 7 & 11 \end{bmatrix}$ E. Not possible

7-5. $A = \begin{bmatrix} 1 & 2 & 0 & 1 \\ 3 & 5 & 8 & 2 \\ 1 & 1 & 0 & 2 \\ 0 & 1 & 4 & 1 \end{bmatrix}$ If $a_{i,j} = 8$, what are the possible values of j?

A. 2 B. 2, 3 C. 3

D. 1, 2, 3 E. 2, 3, 4

7-6. Perform the multiplication $\begin{bmatrix} 10 & 1 \\ 2 & 4 \end{bmatrix}\begin{bmatrix} 3 & 0 \\ 1 & 4 \end{bmatrix}$, if possible.

A. $\begin{bmatrix} 31 & 4 \\ 41 & 16 \end{bmatrix}$ B. $\begin{bmatrix} 31 & 14 \\ 41 & 4 \end{bmatrix}$ C. $\begin{bmatrix} 31 & 8 \\ 10 & 4 \end{bmatrix}$

D. $\begin{bmatrix} 31 & 4 \\ 10 & 16 \end{bmatrix}$ E. Not possible

7-7. If A is a matrix of size 7×3 and B is a matrix of size 3×7, what size is AB, if it exists?

A. 7×7 B. 3×3 C. 7×3

D. 3×7 E. Does not exist.

7-8. A is a matrix of size 7×3, B of size 3×7, and C of size 3×3. What is the size of $2AB - 3C$, if it exists.

A. 7×7 B. 7×3 C. 3×7

D. 3×3 E. Does not exist.

7-9. If $A = \begin{bmatrix} 9 & 3 & 1 \\ 1 & 2 & 1 \\ 1 & 0 & 4 \end{bmatrix}$, what is I such that AI = A?

A. $\begin{bmatrix} 0 & 1 \\ 1 & 0 \end{bmatrix}$ B. $\begin{bmatrix} 1 & 0 \\ 0 & 1 \end{bmatrix}$ C. $\begin{bmatrix} 1 & 0 & 0 \\ 0 & 1 & 0 \\ 0 & 0 & 1 \end{bmatrix}$

D. $\begin{bmatrix} 1 & 0 & 1 \\ 0 & 1 & 0 \\ 1 & 0 & 1 \end{bmatrix}$ E. $\begin{bmatrix} 0 & 0 & 1 \\ 0 & 1 & 0 \\ 1 & 0 & 0 \end{bmatrix}$

7-10. Determine the inverse of $\begin{bmatrix} 1 & 5 \\ 0 & 1 \end{bmatrix}$, if it exists.

A. $\begin{bmatrix} \frac{1}{5} & -\frac{2}{5} \\ 0 & \frac{1}{5} \end{bmatrix}$ B. $\begin{bmatrix} \frac{1}{5} & \frac{2}{5} \\ 0 & \frac{1}{5} \end{bmatrix}$ C. $\begin{bmatrix} 1 & -5 \\ 0 & 1 \end{bmatrix}$

D. $\begin{bmatrix} 0 & -5 \\ 10 & 1 \end{bmatrix}$ E. Does not exit.

7-11. Determine the inverse of $\begin{bmatrix} 1 & 4 & 2 \\ 6 & 1 & 1 \\ 0 & 1 & 2 \end{bmatrix}$, if it exists.

A. $\begin{bmatrix} .714 & .642 & .112 \\ .413 & .281 & .934 \\ .714 & 0 & .142 \end{bmatrix}$ B. $\begin{bmatrix} -.029 & .171 & -.057 \\ .343 & -.057 & -.314 \\ -.171 & .029 & .657 \end{bmatrix}$

C. $\begin{bmatrix} .412 & .413 & .912 \\ .815 & .217 & .716 \\ 0 & .922 & .721 \end{bmatrix}$ D. $\begin{bmatrix} .512 & .413 & .814 \\ .799 & .211 & .333 \\ 0 & .922 & .369 \end{bmatrix}$

E. Does not exist.

7-12. Solve the system of equations using the matrix inverse method.

$$3x+4y=17$$
$$2x+5y=16$$

A. (3, 4) B. (17, 16) C. (3, 5)

D. (3, 2) E. Does not exist.

7-13. Solve the system of equations using the matrix inverse method.

$$4x+3y+2z=20$$
$$4x-y+3z=13$$
$$5x+y-6z=11$$

A. (3, 2, 1) B. (1, 0, 3) C. (2, 1, 0)

D. (2, −1, 2) E. Does not exist.

7-14. Consider the economy defined by the input-output matrix A and open sector D. Determine the output levels required of the industries to meet the demands of the industries and the open sector.

$$A=\begin{bmatrix} .2 & .5 \\ .4 & .1 \end{bmatrix},\ D=\begin{bmatrix} 200 \\ 280 \end{bmatrix}$$

A. $\begin{bmatrix} 20 \\ -172 \end{bmatrix}$ B. $\begin{bmatrix} 200 \\ 280 \end{bmatrix}$ C. $\begin{bmatrix} 180 \\ -60 \end{bmatrix}$

D. $\begin{bmatrix} 76 \\ 276 \end{bmatrix}$ E. $\begin{bmatrix} 615 \\ 585 \end{bmatrix}$

7-15. Compute the determinant $\begin{vmatrix} 6 & 2 \\ 5 & 3 \end{vmatrix}$.

A. 9 B. −9 C. 8 D. −8 E. 28

7-16. Compute the determinant $\begin{vmatrix} 5 & -5 & 5 \\ 1 & 2 & 3 \\ 4 & 5 & 6 \end{vmatrix}$.

A. −12 B. 0 C. 12 D. −60 E. 60

7-17. Compute the minor $a_{3,1}$ of A = $\begin{vmatrix} 4 & 8 & 4 \\ 4 & 8 & 4 \\ 4 & 4 & 8 \end{vmatrix}$.

A. 4 B. 8 C. 0 D. −4 E. −8

7-18. Solve the system using Cramer's rule if it is applicable. If Cramer's rule cannot be used say so.

$$3x+2y=17$$
$$7x-2y=13$$

A. (5, 11) B. (14, 0.5) C. (3, 4)

D. (3, 2) E. Cramer's rule cannot be used.

7-19. Solve the system using Cramer's rule if it is applicable. If Cramer's rule cannot be used say so.

$$7x-4y=7$$
$$-14x+8y=2$$

A. (1, −1) B. (3, 4) C. (1, 1)

D. (−1, 1) E. Cramer's rule cannot be used.

7-20. Solve the system using Cramer's rule if it is applicable. If Cramer's rule cannot be used say so.

$$4x+3y+z=5$$
$$x+y-2z=5$$
$$9x+7y=15$$

A. (1, 2, −1) B. (1, 2, 0) C. (2, 3, 0)

D. (4, 1, 0) E. Cramer's rule cannot be used.

7-21. Compute the determinant of the matrix $\begin{bmatrix} 6 & 4 \\ 5 & 3 \end{bmatrix}$.

A. 2 B. 30 C. –30 D. –2 E. 0

7-22. compute the determinant of the matrix $\begin{bmatrix} 9 & 9 & 9 \\ 2 & 3 & 4 \\ 5 & 6 & 7 \end{bmatrix}$.

A. 0 B. 9 C. –9 D. –1 E. 1

7-23. $A = \begin{bmatrix} 8 & 4 \\ 2 & 5 \end{bmatrix}$. Determine the transpose of A.

A. $\begin{bmatrix} 5 & 4 \\ 2 & 8 \end{bmatrix}$ B. $\begin{bmatrix} 5 & 2 \\ 4 & 8 \end{bmatrix}$ C. $\begin{bmatrix} 2 & 5 \\ 8 & 4 \end{bmatrix}$

D. $\begin{bmatrix} 8 & 4 \\ 2 & 5 \end{bmatrix}$ E. $\begin{bmatrix} 8 & 2 \\ 4 & 5 \end{bmatrix}$

7-24. The matrix $A = \begin{bmatrix} 8 & 2 & 7 \\ 2 & 4 & * \\ 7 & 6 & 2 \end{bmatrix}$ is to be symmetric. Determine the element indicated *.

A. 7 B. 6 C. 2 D. 8 E. 5

7-25. Compute $g_{3,3}$ of the matrix $G = AA'$ where $A = \begin{bmatrix} 1 & 1 & 0 \\ 0 & 0 & 1 \\ 0 & 1 & 1 \end{bmatrix}$.

A. 2 B. 1 C. 0 D. 9 E. 3

Test form 7D Date ________ Name ______________________
Course ______ Sec ______ Student Number ______________

Chapter 7

7-1. Give the size of the matrix $\begin{bmatrix} 9 & 2 & 1 \\ 6 & 3 & 5 \end{bmatrix}$.

7-2. For $A = \begin{bmatrix} -6 & -8 & -9 \\ 7 & -3 & -7 \\ 7 & 3 & 7 \end{bmatrix}$, find $a_{3,2}$.

7-3. Solve the matrix equation $\begin{bmatrix} x & x+y \\ y+z & z \end{bmatrix} = \begin{bmatrix} 1 & 4 \\ 11 & 8 \end{bmatrix}$.

7-4. $A = \begin{bmatrix} 7 & 1 \\ 2 & 3 \end{bmatrix}$, $B = \begin{bmatrix} 4 & 0 \\ 1 & 2 \end{bmatrix}$ Perform the operations A + 2B, if possible.

7-5. $A = \begin{bmatrix} 1 & 2 & 0 & 1 \\ 3 & 5 & 10 & 2 \\ 1 & 1 & 0 & 2 \\ 0 & 1 & 4 & 1 \end{bmatrix}$ If $a_{i,j} = 10$, what are the possible values of j?

7-6. Perform the multiplication $\begin{bmatrix} 2 & 7 \\ 1 & 3 \end{bmatrix}\begin{bmatrix} 1 & 4 \\ 0 & 2 \end{bmatrix}$, if possible.

7-7. If A is a matrix of size 2×6 and B is a matrix of size 6×2, what size is AB, if it exists?

7-8. A is a matrix of size 2×6, B of size 6×2, and C of size 6×6. What is the size of $2AB - 3C$, if it exists.

7-9. If $A = \begin{vmatrix} 7 & 7 & 2 \\ 6 & 1 & 3 \\ 5 & 5 & 4 \end{vmatrix}$, what is I such that AI = A?

7-10. Determine the inverse of $\begin{bmatrix} 1 & 7 \\ 0 & 1 \end{bmatrix}$, if it exists.

7-11. Determine the inverse of $\begin{bmatrix} 1 & 5 & 6 \\ 9 & 3 & 1 \\ 2 & 1 & 0 \end{bmatrix}$, if it exists.

7-12. Solve the system of equations using the matrix inverse method.

$$4x + 2y = 16$$
$$4x + 7y = 26$$

7-13. Solve the system of equations using the matrix inverse method.

$$5x + 2y + 4z = 23$$
$$4x - y + 3z = 13$$
$$5x + y - 6z = 11$$

7-14. Consider the economy defined by the input-output matrix A and open sector D. Determine the output levels required of the industries to meet the demands of the industries and the open sector.

$$A = \begin{bmatrix} .2 & .5 \\ .4 & .1 \end{bmatrix}, D = \begin{bmatrix} 315 \\ 400 \end{bmatrix}$$

7-15. Compute the determinant $\begin{vmatrix} 7 & 2 \\ 5 & 3 \end{vmatrix}$

7-16. Compute the determinant $\begin{vmatrix} 1 & 4 & 2 \\ 6 & 1 & 3 \\ 2 & 1 & 7 \end{vmatrix}$

7-17. Compute the minor $a_{3,1}$ of $A = \begin{vmatrix} 5 & 2 & 5 \\ 5 & 2 & 5 \\ 5 & 5 & 2 \end{vmatrix}$.

7-18. Solve the system using Cramer's rule if it is applicable. If Cramer's rule cannot be used say so.

$$5x + 3y = 27$$
$$7x - 2y = 13$$

7-19. Solve the system using Cramer's rule if it is applicable. If Cramer's rule cannot be used say so.

$$3x - 2y = 7$$
$$-6x + 4y = 5$$

7-20. Solve the system using Cramer's rule if it is applicable. If Cramer's rule cannot be used say so.

$$5x + 2y + z = 1$$
$$x + y - 2z = 5$$
$$11x + 5y = 7$$

7-21. Compute the determinant of the matrix $\begin{bmatrix} 5 & 7 \\ 2 & 4 \end{bmatrix}$.

7-22. Compute the determinant of the matrix $\begin{bmatrix} 4 & 4 & 4 \\ 2 & 3 & 4 \\ 5 & 6 & 7 \end{bmatrix}$.

7-23. $A = \begin{bmatrix} 6 & 2 \\ 5 & 7 \end{bmatrix}$. Determine the transpose of A.

7-24. The matrix $A = \begin{bmatrix} 5 & 2 & 8 \\ 2 & 4 & * \\ 8 & 7 & 2 \end{bmatrix}$ is to be symmetric. Determine the element indicated *.

7-25. Compute $g_{2,1}$ of the matrix $G = AA'$ where $A = \begin{bmatrix} 1 & 1 & 0 \\ 0 & 0 & 1 \\ 0 & 1 & 1 \end{bmatrix}$.

Test form 7E Date ________ Name ________________________
Course ______ Sec ______ Student Number ______________

Chapter 7

7-1. Give the size of the matrix $\begin{bmatrix} 8 & 1 & 3 \\ 6 & 8 & 5 \end{bmatrix}$.

7-2. For A = $\begin{bmatrix} -6 & -8 & -9 \\ 6 & -4 & -6 \\ 6 & 4 & 6 \end{bmatrix}$, find $a_{3,2}$.

7-3. Solve the matrix equation $\begin{bmatrix} x & x+y \\ y+z & z \end{bmatrix} = \begin{bmatrix} 3 & 4 \\ 8 & 7 \end{bmatrix}$.

7-4. $A = \begin{bmatrix} 1 & 4 \\ 2 & 1 \end{bmatrix}$, $B = \begin{bmatrix} 4 & 0 \\ 1 & 2 \end{bmatrix}$ Perform the operations A + 2B, if possible.

7-5. $A = \begin{bmatrix} 1 & 2 & 0 & 1 \\ 3 & 5 & 14 & 2 \\ 1 & 1 & 0 & 2 \\ 0 & 1 & 4 & 1 \end{bmatrix}$ If $a_{i,j} = 14$, what are the possible values of j?

7-6. Perform the multiplication $\begin{bmatrix} 2 & 3 \\ 4 & 1 \end{bmatrix}\begin{bmatrix} 0 & 1 \\ 5 & 3 \end{bmatrix}$, if possible.

7-7. If A is a matrix of size 5×4 and B is a matrix of size 4×5, what size is AB, if it exists?

7-8. A is a matrix of size 5×4, B of size 4×5, and C of size 4×4. What is the size of 2AB – 3C, if it exists.

7-9. If $A = \begin{bmatrix} 8 & 3 & 5 \\ 2 & 3 & 3 \\ 0 & 5 & 1 \end{bmatrix}$, what is I such that AI = A?

7-10. Determine the inverse of $\begin{bmatrix} 1 & 9 \\ 0 & 1 \end{bmatrix}$, if it exists.

7-11. Determine the inverse of $\begin{bmatrix} 2 & 6 & 1 \\ 0 & 2 & 4 \\ 3 & 4 & 9 \end{bmatrix}$, if it exists.

7-12. Solve the system of equations using the matrix inverse method.

$$6x + 2y = 16$$
$$7x + y = 12$$

7-13. Solve the system of equations using the matrix inverse method.

$$2x + 3y + 4z = 16$$
$$4x - y + 3z = 13$$
$$5x + y - 6z = 11$$

7-14. Consider the economy defined by the input-output matrix A and open sector D. Determine the output levels required of the industries to meet the demands of the industries and the open sector.

$$A = \begin{bmatrix} .2 & .5 \\ .4 & .1 \end{bmatrix}, D = \begin{bmatrix} 218 \\ 312 \end{bmatrix}$$

7-15. Compute the determinant $\begin{vmatrix} 8 & 3 \\ 2 & 4 \end{vmatrix}$.

7-16. Compute the determinant $\begin{vmatrix} 2 & 5 & 1 \\ 3 & 7 & 2 \\ 1 & 4 & 1 \end{vmatrix}$.

7-17. Compute the minor $a_{3,1}$ of $A = \begin{vmatrix} 7 & 3 & 7 \\ 7 & 3 & 7 \\ 7 & 7 & 3 \end{vmatrix}$.

7-18. Solve the system using Cramer's rule if it is applicable. If Cramer's rule cannot be used say so.

$$6x+2y=26$$
$$7x-2y=13$$

7-19. Solve the system using Cramer's rule if it is applicable. If Cramer's rule cannot be used say so.

$$5x-3y=7$$
$$-10x+6y=8$$

7-20. Solve the system using Cramer's rule if it is applicable. If Cramer's rule cannot be used say so.

$$2x+5y+z=3$$
$$x+y-2z=5$$
$$5x+11y=11$$

7-21. Compute the determinant of the matrix $\begin{bmatrix} 8 & 3 \\ 2 & 5 \end{bmatrix}$.

7-22. Compute the determinant of the matrix $\begin{bmatrix} 3 & 3 & 3 \\ 2 & 3 & 4 \\ 5 & 6 & 7 \end{bmatrix}$.

7-23. $A = \begin{bmatrix} 9 & 8 \\ 7 & 4 \end{bmatrix}$. Determine the transpose of A.

7-24. The matrix $A = \begin{bmatrix} 5 & 2 & 9 \\ 2 & 4 & * \\ 9 & 8 & 1 \end{bmatrix}$ is to be symmetric. Determine the element indicated *.

7-25. Compute $g_{2,2}$ of the matrix $G = AA'$ where $A = \begin{bmatrix} 1 & 1 & 0 \\ 0 & 0 & 1 \\ 0 & 1 & 1 \end{bmatrix}$.

Test form 7F Date ________ Name ______________________
Course ______ Sec ______ Student Number ______________________

Chapter 7

7-1. Give the size of the matrix $\begin{bmatrix} 6 & 2 & 7 \\ 4 & 3 & 4 \end{bmatrix}$.

7-2. For $A = \begin{bmatrix} -6 & -8 & -9 \\ 5 & -1 & -5 \\ 5 & 1 & 5 \end{bmatrix}$, find $a_{3,2}$.

7-3. Solve the matrix equation $\begin{bmatrix} x & x+y \\ y+z & z \end{bmatrix} = \begin{bmatrix} 7 & 9 \\ 3 & 1 \end{bmatrix}$.

7-4. $A = \begin{bmatrix} 1 & 3 \\ 7 & 2 \end{bmatrix}$, $B = \begin{bmatrix} 4 & 0 \\ 1 & 2 \end{bmatrix}$ Perform the operations A + 2B, if possible.

7-5. $A = \begin{bmatrix} 1 & 2 & 0 & 1 \\ 3 & 5 & 20 & 2 \\ 1 & 1 & 0 & 2 \\ 0 & 1 & 4 & 1 \end{bmatrix}$ If $a_{i,j} = 20$, what are the possible values of j?

7-6. Perform the multiplication $\begin{bmatrix} 8 & 9 \\ 7 & 5 \end{bmatrix}\begin{bmatrix} 3 & 1 \\ 0 & 4 \end{bmatrix}$, if possible.

7-7. If A is a matrix of size 5×4 and B is a matrix of size 4×5, what size is AB, if it exists?

7-8. A is a matrix of size 5×7, B of size 7×5, and C of size 7×7. What is the size of $2AB - 3C$, if it exists.

7-9. If $A = \begin{bmatrix} 8 & 3 & 5 \\ 6 & 1 & 4 \\ 2 & 2 & 1 \end{bmatrix}$, what is I such that AI = A?

7-10. Determine the inverse of $\begin{bmatrix} 1 & 12 \\ 0 & 1 \end{bmatrix}$, if it exists.

7-11. Determine the inverse of $\begin{bmatrix} 9 & 1 & 4 \\ 2 & 0 & 1 \\ 3 & 3 & 7 \end{bmatrix}$, if it exists.

7-12. Solve the system of equations using the matrix inverse method.

$$5x + 4y = 17$$
$$2x + 4y = 14$$

7-13. Solve the system of equations using the matrix inverse method.

$$7x + 5y + z = 32$$
$$4x - y + 3z = 13$$
$$5x + y - 6z = 11$$

7-14. Consider the economy defined by the input-output matrix A and open sector D. Determine the output levels required of the industries to meet the demands of the industries and the open sector.

$$A = \begin{bmatrix} .2 & .5 \\ .4 & .1 \end{bmatrix}, D = \begin{bmatrix} 112 \\ 300 \end{bmatrix}$$

7-15. Compute the determinant $\begin{vmatrix} 8 & 4 \\ 7 & 9 \end{vmatrix}$

7-16. Compute the determinant $\begin{vmatrix} 5 & 1 & 2 \\ 7 & 3 & 4 \\ 1 & 1 & 7 \end{vmatrix}$.

7-17. Compute the minor $a_{3,1}$ of $A = \begin{vmatrix} 8 & 6 & 8 \\ 8 & 6 & 8 \\ 8 & 8 & 6 \end{vmatrix}$.

7-18. Solve the system using Cramer's rule if it is applicable. If Cramer's rule cannot be used say so.

$$2x+7y=34$$
$$7x-2y=13$$

7-19. Solve the system using Cramer's rule if it is applicable. If Cramer's rule cannot be used say so.

$$4x-7y=5$$
$$-8x+14y=7$$

7-20. Solve the system using Cramer's rule if it is applicable. If Cramer's rule cannot be used say so.

$$6x+2y+z=5$$
$$x+y-2z=5$$
$$13x+5y=15$$

7-21. Compute the determinant of the matrix $\begin{bmatrix} 7 & 2 \\ 4 & 5 \end{bmatrix}$.

7-22. Compute the determinant of the matrix $\begin{bmatrix} 7 & 7 & 7 \\ 2 & 3 & 4 \\ 5 & 6 & 7 \end{bmatrix}$.

7-23. $A = \begin{bmatrix} 4 & 7 \\ 3 & 6 \end{bmatrix}$. Determine the transpose of A.

7-24. The matrix $A = \begin{bmatrix} 5 & 2 & 2 \\ 2 & 4 & * \\ 2 & 7 & 3 \end{bmatrix}$ is to be symmetric. Determine the element indicated *.

7-25. Compute $g_{2,3}$ of the matrix $G = AA'$ where $A = \begin{bmatrix} 1 & 1 & 0 \\ 0 & 0 & 1 \\ 0 & 1 & 1 \end{bmatrix}$.

Test form 8A Date _________ Name _________________________
Course ______ Sec ______ Student Number ________________

Chapter 8

8-1. Determine the first four terms of the arithmetic sequence $a_n = 4n + 7$.

A. 1, 2, 3, 4 B. 7, 14, 21, 28

C. 8, 9, 10, 11 D. 11, 15, 19, 23

E. 0, 11, 15, 19

8-2. $a_n = 5n^2 + 6$. Find the fifth term of the sequence.

A. 86 B. 131 C. 155 D. 29 E. 104

8-3. If $a_n = a_{n-1} + 7$ and $a_1 = 5$, give the first 5 terms of the sequence.

A. 5, 12, 19, 26, 33 B. 12, 19, 26, 33, 40

C. 5, 5, 5, 5, 5 D. 7, 14, 21, 28, 35

E. 0, 7, 14, 21, 28

8-4. Evaluate $\sum_{k=1}^{5}(3k+6)$.

A. 50 B. 75 C. 51 D. 90 E. 56

8-5. Express $5+6+7+8$ using sigma notation.

A. $\sum_{k=1}^{4}(k^2+1)$ B. $\sum_{k=1}^{4}(k^2+k)$

C. $\sum_{k=0}^{3}(k^2+k)$ D. $\sum_{k=0}^{3}(4+k)$

E. $\sum_{k=1}^{4}(4+k)$

8-6. Find the partial sum, S_3, of the sequence $a_k = 4k - 2$.

A. 18 B. 30 C. 10 D. 14 E. 22

8-7. Determine the mean of the first 3 terms of the sequence $a_k = 8k - 5$.

A. 8 B. 9 C. 10 D. 11 E. 12

8-8. 4, 7, 10, 13, 16 are the beginning of an arithmetic sequence. Find the common difference.

A. 4 B. 3 C. -3 D. 0 E. 1

8-9. Give the next two terms of the sequence 0, 2, 4, 6.

A. 8, 16 B. 2, 0 C. $-8, -10$

D. 8, 10 E. 6, 8

8-10. The first term $a_1 = 5$ and the common difference $d = 4$ of an arithmetic sequence are given. Find the first four terms.

A. 4, 4, 4, 4 B. 5, 5, 5, 5

C. 4, 9, 14, 19 D. 5, 9, 13, 17

E. 9, 13, 17, 21

8-11. The first term of an arithmetic sequence is $a_1 = 6$ and the difference is $d = 3$. Find the 7th term.

A. 18 B. 21 C. 24 D. 27 E. 30

8-12. The second term of an arithmetic sequence is 9 and the third term is 12. Find the 7th term.

A. 18 B. 21 C. 24 D. 27 E. 30

8-13. The third term of an arithmetic sequence is 12 and the sixth is 21. Find the 7th term.

A. 18 B. 21 C. 24 D. 27 E. 30

8-14. Determine the 5th partial sum of the sequence where $a_1 = 6$ and $d = 3$.

A. 54 B. 63 C. 66 D. 60 E. 69

8-15. Find the common ratio for the geometric sequence 1, –2, 4, –8,....

A. 2 B. –2 C. $\frac{1}{2}$ D. $-\frac{1}{2}$ E. 1

8-16. The first term and the common ratio of a geometric sequence are $a_1 = 4$ and $r = 3$. Find the 8th term.

A. $4(3)^8$ B. $4(3)^9$ C. $4(3)^7$ D. $(12)^7$ E. $(12)^8$

8-17. Find the 10th partial sum for the sequence $a_n = 4\left(\frac{1}{3}\right)^{n-1}$.

A. 5.9999 B. 5.9998 C. 6.0002

D. 5.9996 E. 6.0004

8-18. Find the sum of the infinite geometric series having first term $a_1 = 5$ and common ratio 1/4.

A. 6.6667 B. 6.6669 C. 6.6711

D. 6.6713 E. 6.6719

8-19. Write $0.4\overline{41}$ as a fraction.

A. $\frac{441}{999}$ B. $\frac{41}{1000}$ C. $\frac{41}{100}$

D. $\frac{41}{100}$ E. $\frac{437}{990}$

8-20. Mathematical induction can be used to show $3+6+9+\ldots+a_n =$

A. $n(n+1)$ B. $n(n-1)$ C. $\frac{3}{2}n(n-1)$

D. $\frac{3}{2}n(n+1)$ E. $3n(n+1)$

8-21. Expand $(a+3b)^5$ using Pascal's triangle.

A. $a^5 + 3a^4b + 9a^3b^2 + 27a^2b^3 + 81ab^4 + 243b^5$

B. $a^5 + 15a^4b + 90a^3b^2 + 270a^2b^3 + 405ab^4 + 243b^5$

C. $a^5 + 15a^4b + 135a^3b^2 + 405a^2b^3 + 405ab^4 + 243b^5$

D. $a^5 + 5a^4b + 10a^3b^2 + 10a^2b^3 + 5ab^4 + b^5$

E. $3(a^5 + 5a^4b + 10a^3b^2 + 10a^2b^3 + 5ab^4 + b^5)$

8-22. $\frac{90!}{86!} =$

A. 61,324,560 B. 5,212,587600 C. 704,800

D. 7.920 E. 90

8-23. $\binom{6}{4} =$

A. 34 B. 20 C. 720 D. 360 E. 15

8-24. Determine the 3rd term of $(x + y)^4$.

A. $6x^3y^z$ B. $6x^2y^2$ C. $4x^2y^2$

D. $4xy^3$ E. $4xy$

8-25. Determine the 3rd term of $(x - y)^4$.

A. $-6x^3y^z$ B. $6x^2y^2$ C. $-4x^2y^2$

D. $-4xy^3$ E. $-4xy$

Test form 8B　　Date ________　　Name ______________________
Course ______ Sec ______　　Student Number ______________

Chapter 8

8-1. Determine the first four terms of the arithmetic sequence $a_n = 4n + 10$.

A. 1, 2, 3, 4　　B. 14, 18, 22, 26
C. 0, 14, 22, 26　　D. 10, 20, 30, 40
E. 0, 11, 12, 13

8-2. $a_n = 4n^2 + 3$. Find the fifth term of the sequence.

A. 103　B. 112　C. 23　D. 67　E. 76

8-3. If $a_n = a_{n-1} + 8$ and $a_1 = 5$, give the first 5 terms of the sequence.

A. 5, 13, 21, 29, 37　　B. 13, 21, 29, 37, 45
C. 5, 5, 5, 5, 5　　D. 8, 16, 24, 32, 40
E. 0, 8, 16, 24, 32

8-4. Evaluate $\sum_{k=1}^{5}(4k + 2)$.

A. 70　B. 62　C. 90　D. 60　E. 54

8-5. Express $10 + 11 + 12 + 13$ using sigma notation.

A. $\sum_{k=1}^{4}(9 + k)$　　B. $\sum_{k=0}^{3}(9 + k)$

C. $\sum_{k=1}^{4}(k^2 + 1)$　　D. $\sum_{k=0}^{3}k^2 + k$

E. $\sum_{k=1}^{4}k^2 + k$

8-6. Find the partial sum, S_3, of the sequence $a_k = 4k - 5$.

A. 15　B. 17　C. 9　D. 33　E. 21

8-7. Determine the mean of the first 3 terms of the sequence $a_k = 9k - 5$.

A. 13 B. 14 C. 15 D. 16 E. 17

8-8. 5, 8, 11, 14, 17 are the beginning of an arithmetic sequence. Find the common difference.

A. 3 B. 5 C. −3 D. 0 E. 1

8-9. Give the next two terms of the sequence 0, 3, 6, 9.

A. 12, 15 B. 12, 21 C. 3, 0

D. −12, −15 E. 9, 12

8-10. The first term $a_1 = 6$ and the common difference $d = 3$ of an arithmetic sequence are given. Find the first four terms.

A. 3, 3, 3, 3 B. 6, 6, 6, 6

C. 6, 9, 12, 15 D. 3, 9, 15, 21

E. 9, 12, 15, 18

8-11. The first term of an arithmetic sequence is $a_1 = 6$ and the difference is $d = 3$. Find the 5th term.

A. 15 B. 18 C. 21 D. 24 E. 28

8-12. The second term of an arithmetic sequence is 9 and the third term is 12. Find the 5th term.

A. 15 B. 18 C. 21 D. 24 E. 28

8-13. The third term of an arithmetic sequence is 12 and the sixth is 21. Find the 5th term.

A. 15 B. 18 C. 21 D. 24 E. 28

8-14. Determine the 7th partial sum of the sequence where $a_1 = 6$ and $d = 3$.

A. 104 B. 108 C. 110 D. 105 E. 120

8-15. Find the common ratio for the geometric sequence 3, –6, 12, –24....

A. 2 B. –2 C. $\frac{1}{2}$ D. $-\frac{1}{2}$ E. 3

8-16. The first term and the common ratio of a geometric sequence are $a_1 = 4$ and $r = 3$. Find the 10th term.

A. $4(3)^{10}$ B. $4(3)^{11}$ C. $4(3)^9$ D. $(12)^9$ E. $(12)^{10}$

8-17. Find the 5th partial sum for the sequence $a_n = 6\left(\frac{1}{2}\right)^{n-1}$.

A. 11.625 B. 11.630 C. 11.641

D. 11.644 E. 11.652

8-18. Find the sum of the infinite geometric series having first term $a_1 = 6$ and common ratio 1/5.

A. 7.4982 B. 7.5000 C. 7.5011

D. 7.5014 E. 7.5025

8-19. Write $0.5\overline{41}$ as a fraction.

A. $\frac{541}{999}$ B. $\frac{536}{990}$ C. $\frac{41}{99}$

D. $\frac{41}{100}$ E. $\frac{541}{1000}$

8-20. Mathematical induction can be used to show $4+8+12+\ldots+a_n =$

A. $n(n+1)$ B. $n(n-1)$ C. $\frac{4}{2}n(n-1)$

D. $\frac{4}{2}n(n+1)$ E. $4n(n+1)$

8-21. Expand $(3a+b)^5$ using Pascal's triangle.

A. $243a^5 + 81a^4b + 27a^3b^2 + 9a^2b^3 + 3ab^4 + b^5$

B. $243a^5+405a^4b+270a^3b^2+90a^2b^3+15ab^4+b^5$

C. $243^5+405a^4b+405a^3b^2+135a^2b^3+15ab^4+b^5$

D. $a^5+5a^4b+10a^3b^2+10a^2b^3+5ab^4+b^5$

E. $3(a^5+5a^4b+10a^3b^2+10a^2b^3+5ab^4+b^5)$

8-22. $\frac{90!}{88!}=$

A. 61,324,560 B. 5,212,587,600 C. 704,800

D. 8,010 E. 90

8-23. $\binom{6}{2}=$

A. 36 B. 32 C. 15 D. 720 E. 30

8-24. Determine the 2nd term of $(x+y)^4$.

A. $6x^3y$ B. $6x^2y^2$ C. $4x^3y$

D. $4xy^3$ E. $4xy$

8-25. Determine the 2nd term of $(x-y)^4$.

A. $6x^3y$ B. $6x^2y^2$ C. $-4x^3y$

D. $4xy^3$ E. $4xy$

Test form 8C　　Date ________　　Name ______________________

Course ______ Sec ______　　Student Number ______________

Chapter 8

8-1. Determine the first four terms of the arithmetic sequence $a_n = 4n + 8$.

A. 1, 2, 3, 4　　B. 8, 16, 24, 32

C. 9, 10, 11, 12　　D. 0, 12, 16, 20

E. 12, 16, 20, 24

8-2. $a_n = 3n^2 + 7$. Find the fifth term of the sequence.

A. 55　B. 76　C. 22　D. 110　E. 82

8-3. If $a_n = a_{n-1} + 10$ and $a_1 = 5$, give the first 5 terms of the sequence.

A. 5, 15, 25, 35, 45　　B. 15, 25, 35, 45, 55

C. 5, 5, 5, 5, 5　　D. 10, 20, 30, 40, 50

E. 0, 10, 20, 30, 40

8-4. Evaluate $\sum_{k=1}^{5}(2k+9)$.

A. 70　B. 40　C. 35　D. 75　E. 39

8-5. Express $26 + 27 + 28 + 29$ using sigma notation.

A. $\sum_{k=1}^{4} 25$　　B. $\sum_{k=1}^{4}(25+k)$

C. $\sum_{k=1}^{4}(k^2+k)$　　D. $\sum_{k=0}^{3}(25+k)$

E. $\sum_{k=0}^{3}(k^2+k)$

8-6. Find the partial sum, S_3, of the sequence $a_k = 5k - 2$.

A. 36　B. 24　C. 32　D. 17　E. 13

8-7. Determine the mean of the first 3 terms of the sequence $a_k = 7k - 5$.

A. 10 B. 9 C. 8 D. 11 E. 12

8-8. 8, 11, 14, 17, 20 are the beginning of an arithmetic sequence. Find the common difference.

A. 8 B. 1 C. –3 D. 0 E. 3

8-9. Give the next two terms of the sequence 0, 5, 10, 15.

A. 20, 35 B. 20, 25 C. 5, 0

D. –20, 25 E. 15, 20

8-10. The first term $a_1 = 3$ and the common difference $d = 4$ of an arithmetic sequence are given. Find the first four terms.

A. 3, 3, 3, 3 B. 4, 4, 4, 4

C. 3, 7, 11, 15 D. 4, 7, 11, 15

E. 7, 11, 15, 19

8-11. The first term of an arithmetic sequence is $a_1 = 6$ and the difference is $d = 3$. Find the 8th term.

A. 18 B. 21 C. 24 D. 27 E. 30

8-12. The second term of an arithmetic sequence is 9 and the third term is 12. Find the 8th term.

A. 18 B. 21 C. 24 D. 27 E. 30

8-13. The third term of an arithmetic sequence is 12 and the sixth is 21. Find the 8th term.

A. 18 B. 21 C. 24 D. 27 E. 30

8-14. Determine the 8th partial sum of the sequence where $a_1 = 6$ and $d = 3$.

A. 102 B. 118 C. 124 D. 130 E. 132

8-15. Find the common ratio for the geometric sequence 4, −8, 16, −32,....

A. 2 B. −2 C. $\frac{1}{2}$ D. $-\frac{1}{2}$ E. 4

8-16. The first term and the common ratio of a geometric sequence are $a_1 = 4$ and $r = 3$. Find the 12th term.

A. $4(3)^{12}$ B. $4(3)^{13}$ C. $4(3)^{11}$ D. $(12)^{11}$ E. $(12)^{12}$

8-17. Find the 6th partial sum for the sequence $a_n = 8\left(\frac{1}{4}\right)^{n-1}$.

A. 10.6637 B. 10.6641 C. 10.6645

D. 10.6653 E. 10.6659

8-18. Find the sum of the infinite geometric series having first term $a_1 = 8$ and common ratio 1/3.

A. 11.9854 B. 11.9892 C. 11.9972

D. 12.0000 E. 12.0142

8-19. Write $0.6\overline{41}$ as a fraction.

A. $\frac{635}{990}$ B. $\frac{641}{999}$ C. $\frac{41}{99}$

D. $\frac{41}{100}$ E. $\frac{641}{1000}$

8-20. Mathematical induction can be used to show $6+12+18+\ldots+a_n =$

A. $n(n+1)$ B. $n(n-1)$ C. $\frac{6}{2}n(n-1)$

D. $\frac{6}{2}n(n+1)$ E. $6n(n+1)$

8-21. Expand $(a-3b)^5$ using Pascal's triangle.

A. $a^5 - 3a^4b + 9a^3b^2 - 27a^2b^3 + 81ab^4 - 243b^5$

B. $a^5-15a^4b+90a^3b^2-270a^2b^3+405ab^4-243b^5$

C. $a^5-15a^4b+135a^3b^2-405a^2b^3+405ab^4-243b^5$

D. $a^5-5a^4b+10a^3b^2-10a^2b^3+5ab^4-b^5$

E. $3(a^5-5a^4b+10a^3b^2-10a^2b^3+5ab^4-b^5)$

8-22. $\frac{90!}{85!}=$

A. 5,273,912,160 B. 5,212,587,600 C. 704,800

D. 7,920 E. 90

8-23. $\binom{8}{3}=$

A. 56 B. 40,320 C. 55

D. 336 E. 64

8-24. Determine the 4th term of $(x+y)^4$.

A. $6x^3y$ B. $6x^2y^2$ C. $4x^2y^2$

D. $4xy^3$ E. $4xy$

8-25. Determine the 4th term of $(x-y)^4$.

A. $6x^3y$ B. $6x^2y^2$ C. $4x^2y^2$

D. $-4xy^3$ E. $4xy$

Test form 8D Date _________ Name ___________________________
Course ______ Sec ______ Student Number _________________

Chapter 8

8-1. Determine the first four terms of the arithmetic sequence $a_n = 4n + 6$.

8-2. $a_n = 4n^2 + 12$. Find the fifth term of the sequence.

8-3. If $a_n = a_{n-1} + 12$ and $a_1 = 5$, give the first 5 terms of the sequence.

8-4. Evaluate $\sum_{k=1}^{5} (5k + 4)$.

8-5. Express $9 + 10 + 11 + 12$ using sigma notation.

8-6. Find the partial sum, S_3, of the sequence $a_k = 8k - 5$.

8-7. Determine the mean of the first 3 terms of the sequence $a_k = 10k - 3$.

8-8. 7, 12, 17, 22, 27 are the beginning of an arithmetic sequence. Find the common difference.

8-9. Give the next two terms of the sequence 0, 6, 12, 18.

8-10. The first term $a_1 = 9$ and the common difference $d = 3$ of an arithmetic sequence are given. Find the first four terms.

8-11. The first term of an arithmetic sequence is $a_1 = 6$ and the difference is $d - 3$. Find the 9th term.

8-12. The second term of an arithmetic sequence is 9 and the third rerm is 12. Find the 9th term.

8-13. The third term of an arithmetic sequence is 12 and the sixth is 21. Find the 9th term.

8-14. Determine the 10th partial sum of the sequence where $a_1 = 6$ and $d = 3$.

8-15. Find the common ratio for the geometric sequence 10, –20, 40, –80,....

8-16. The first term and the common ratio of a geometric sequence are $a_1 = 4$ and $r = 3$. Find the 15th term.

8-17. Find the 11th partial sum for the sequence $a_n = 5\left(\frac{1}{10}\right)^{n-1}$.

8-18. Find the sum of the infinite geometric series having first term $a_1 = 9$ and common ratio 1/4.

8-19. Write $0.7\overline{41}$ as a fraction.

8-20. Mathematical induction can be used to show $10 + 20 + 30 + \ldots + a_n =$

8-21. Expand $(a + 4b)^5$ using Pascal's triangle.

8-22. $\frac{94!}{92!} =$

8-23. $\binom{9}{2} =$

8-24. Determine the 5th term of $(x + y)^8$.

8-25. Determine the 5th term of $(x - y)^8$.

Test form 8E Date __________ Name ____________________________
Course ______ Sec ______ Student Number __________________

Chapter 8

8-1. Determine the first four terms of the sequence $a_n = 4n + 3$.

8-2. $a_n = 8n^2 + 7$. Find the fifth term of the sequence.

8-3. If $a_n = a_{n-1} + 6$ and $a_1 = 5$, give the first 5 terms of the sequence.

8-4. Evaluate $\sum_{k=1}^{5} (4k + 7)$.

8-5. Express $12 + 13 + 14 + 15$ using sigma notation.

8-6. Find the partial sum, S_3, of the sequence $a_k = 6k - 5$.

8-7. Determine the mean of the first 3 terms of the sequence $a_k = 15k - 5$.

8-8. 9, 15, 21, 27, 33 are the beginning of an arithmetic sequence. Find the common difference.

8-9. Give the next two terms of the sequence 0, 9, 18, 27.

8-10. The first term $a_1 = 5$ and the common difference $d = 8$ of an arithmetic sequence are given. Find the first four terms.

8-11. The first term of an arithmetic sequence is $a_1 - 6$ and the difference is $d = 3$. Find the 11th term.

8-12. The second term of an arithmetic sequence is 9 and the third term is 12. Find the 11th term.

8-13. The third term of an arithmetic sequence is 12 and the sixth is 21. Find the 11th term.

8-14. Determine the 9th partial sum of the sequence where $a_1 = 6$ and $d = 3$.

8-15. Find the common ratio for the geometric sequence $-5, 10, -20, 40, \ldots$.

8-16. The first term and the common ratio of a geometric sequence are $a_1 = 4$ and $r = 3$. Find the 16th term.

8-17. Find the 8th partial sum for the sequence $a_n = 7\left(\frac{1}{9}\right)^{n-1}$.

8-18. Find the sum of the infinite geometric series having first term $a_1 = 10$ and common ratio 1/7.

8-19. Write $0.8\overline{41}$ as a fraction.

8-20. Mathematical induction can be used to show $12 + 24 + 36 + \ldots + a_n =$

8-21. Expand $(a - 4b)^5$ using Pascal's triangle.

8-22. $\frac{100!}{97!} =$

8-23. $\binom{11}{4} =$

8-24. Determine the 3rd term of $(x + y)^8$.

8-25. Determine the 3rd term of $(x - y)^8$.

Test form 8F	Date __________	Name ____________________________
Course ______ Sec ______	Student Number ________________

Chapter 8

8-1. Determine the first four terms of the arithmetic sequence $a_n = 4n + 5$.

8-2. $a_n = 12n^2 + 9$. Find the fifth term of the sequence.

8-3. If $a_n = a_{n-1} + 9$ and $a_1 = 5$, give the first 5 terms of the sequence.

8-4. Evaluate $\sum_{k=1}^{5} (5k + 8)$.

8-5. Express $17 + 18 + 19 + 20$ using sigma notation.

8-6. Find the partial sum, S_3, of the sequence $a_k = 7k - 3$.

8-7. Determine the mean of the first 3 terms of the sequence $a_k = 4k - 5$.

8-8. 4, 9, 14, 19, 24 are the beginning of an arithmetic sequence. Find the common difference.

8-9. Give the next two terms of the sequence 0, 10, 20, 30.

8-10. The first term $a_1 = 7$ and the common difference $d = 4$ of an arithmetic sequence are given. Find the first four terms.

8-11. The first term of an arithmetic sequence is $a_1 = 6$ and the difference is $d = 3$. Find the 13th term.

8-12. The second term of an arithmetic sequence is 9 and the third term is 12. Find the 13th term.

8-13. The third term of an arithmetic sequence is 12 and the sixth is 21. Find the 13th term.

8-14. Determine the 12th partial sum of the sequence where $a_1 = 6$ and $d = 3$.

8-15. Find the common ratio for the geometric sequence –6, 12, –24, 48,....

8-16. The first term and the common ratio of a geometric sequence are $a_1 = 4$ and $r = 3$. Find the 20th term.

8-17. Find the 11th partial sum for the sequence $a_n = 10\left(\frac{1}{12}\right)^{n-1}$.

8-18. Find the sum of the infinite geometric series having first term $a_1 = 9$ and common ratio 1/8.

8-19. Write $0.9\overline{41}$ as a fraction.

8-20. Mathematical induction can be used to show $14 + 28 + 42 + \ldots + a_n =$

8-21. Expand $(4a + b)^5$ using Pascal's triangle.

8-22. $\dfrac{102!}{100!} =$

8-23. $\dbinom{10}{4} =$

8-24. Determine the 2nd term of $(x + y)^9$.

8-25. Determine the 2nd term of $(x - y)^9$.

Test form 9A Date _________ Name ___________________________
Course ______ Sec ______ Student Number ________________

Chapter 9

9-1. 7!

A. 7 B. 49 C. 343 D. 5040 E. 5041

9-2. ${}_7P_2 =$

A. 7 B. 5 C. 42 D. 210 E. $\frac{7}{2}$

9-3. ${}_6C_5 =$

A. 6 B. 5 C. 30 D. 2 E. 15

9-4. A person has 6 colors of paint and 7 colors of floor tile to choose from. If one of each are to be selected, how many color schemes are possible for a room?

A. 38 B. 39 C. 40 D. 41 E. 42

9-5. A person must select one type of drink and one snack at the movies. If there are 7 different drinks and 4 different snacks available, how many possibilities are there?

A. 27 B. 28 C. 29 D. 30 E. 31

9-6. If there are 10 students in a class and each is to select a different special assignment from 10 that are available, in how many ways can the selection be made?

A. 10! B. ${}_{10}P_1$ C. ${}_{10}C_{10}$

D. ${}_{10}C_0$ E. 10

9-7. In how many ways can the top three of 20 teams be awarded prizes in an academic competition?

A. $20!$ B. ${}_{20}P_3$ C. ${}_{20}C_3$

D. ${}_{20}C_0$ E. $\frac{20}{3}$

9-8. In how many ways can a student select 7 of 10 questions on a test?

A. 7^{10} B. 10^7 C. ${}_{10}P_7$ D. ${}_{10}C_7$ E. $\frac{10!}{7!}$

9-9. Of 10 students 7 are to be given identical awards. In how many ways can this be done?

A. 7^{10} B. 7^{10} C. ${}_{10}P_7$ D. ${}_{10}C_7$ E. $\frac{10!}{7!}$

9-10. A 6-sided die is rolled. What is the probability the side up will be 2?

A. 2 B. $\frac{1}{2}$ C. 6 D. $\frac{1}{6}$ E. 0

9-11. A fair coin is tossed 10 times. What is the probability of getting 7 heads?

A. $\frac{1}{2^{10}}$ B. $\frac{{}_{10}P_7}{2^{10}}$ C. $\frac{{}_{10}C_7}{2^{10}}$

D. $\frac{1}{2}$ E. $\frac{7}{10}$

9-12. If one card is selected from a deck of 52 cards, what is the probability it is a 3?

A. $\frac{1}{52}$ B. $\frac{1}{4}$ C. $\frac{1}{13}$ D. $\frac{1}{48}$ E. $\frac{1}{12}$

9-13. Consider the experiment of tossing two dice. Determine the probability that the sum that comes up is 7.

A. $\frac{7}{36}$ B. $\frac{1}{36}$ C. $\frac{6}{36}$ D. $\frac{2}{36}$ E. $\frac{35}{36}$

9-14. Two cards are selected from a deck of 52 cards. What is the probability one is a 9 and the other is an 8?

A. $\frac{4}{52} \cdot \frac{4}{51}$ B. $\frac{1}{52} \cdot \frac{1}{51}$

C. $\frac{4}{52} \cdot \frac{4}{52}$ D. $\frac{4}{52} \cdot \frac{3}{51}$

E. $2 \cdot \frac{4}{52} \cdot \frac{4}{51}$

9-15. A card is selected from a deck of 52 cards. What is the probability of getting a 5 or a spade?

A. $\frac{1}{52}$ B. $\frac{1}{13}$ C. $\frac{1}{4}$ D. $\frac{4}{13}$ E. $\frac{1}{3}$

9-16. A card is selected from a deck of 52 cards. What is the probability of getting a jack and a club?

A. $\frac{1}{13}$ B. $\frac{1}{52}$ C. $\frac{1}{4}$ D. $\frac{4}{13}$ E. $\frac{1}{3}$

9-17. A card is selected from a deck of 52 cards. What is the probability of getting a 10 or an 8?

A. $\frac{1}{52}$ B. $\frac{2}{13}$ C. $\frac{1}{4}$ D. $\frac{1}{13}$ E. $\frac{1}{3}$

9-18. Suppose there are 6 freshmen, 8 sophomores, and 11 juniors in a group. One student is selected at random. What is the probability a junior is selected?

A. $\frac{6}{25}$ B. $\frac{8}{25}$ C. $\frac{11}{25}$ D. $\frac{6}{27}$ E. $\frac{8}{27}$

9-19. What is the probability of being dealt a 5-card hand with 4 jacks?

A. $\frac{4}{52}$ B. $\frac{{}_{48}C_1}{{}_{52}C_5}$ C. $13\frac{{}_{48}C_1}{{}_{52}C_5}$

D. $\frac{{}_{52}C_4}{{}_{52}C_5}$ E. $13\frac{{}_{52}C_4}{{}_{52}C_5}$

9-20. A box contains 20 good radios and 4 defective radios. What is the probability of getting exactly one defective radio if 3 radios are selected?

A. $\frac{{}_{20}C_2 \, {}_4C_1}{{}_{24}C_3}$ B. $\frac{{}_{20}C_3}{{}_{24}C_3}$ C. $\frac{{}_4C_1}{{}_{20}C_3}$

D. $\frac{{}_4C_1}{{}_{24}C_2}$ E. $\frac{{}_4C_1}{{}_{20}C_2}$

9-21. A box contains 20 good radios and 4 defective radios. What is the probability of getting at most one defective radio if 3 radios are selected?

A. $\frac{{}_{20}C_2 \, {}_4C_1}{{}_{24}C_3}$ B. $\frac{{}_{20}C_2 \, {}_4C_1}{{}_{24}C_3} + \frac{{}_{20}C_3}{{}_{24}C_3}$

C. $\frac{{}_4C_1}{{}_{20}C_3}$ D. $\frac{{}_4C_1}{{}_{24}C_2}$

E. $\frac{{}_{20}C_1}{{}_{24}C_2}$

9-22. A box contains 20 good radios and 4 defective radios. What is the probability of getting at least two defective radios if 3 radios are selected?

A. $1 - \frac{{}_{20}C_2 \, {}_4C_1}{{}_{24}C_3} - \frac{{}_{20}C_3}{{}_{24}C_3}$ B. $1 - \frac{{}_{20}C_2 \, {}_4C_1}{{}_{24}C_3}$

C. $\frac{{}_4C_1}{{}_{20}C_3}$ D. $\frac{{}_4C_1}{{}_{24}C_2}$

E. $\frac{{}_{20}C_1}{{}_{24}C_2}$

9-23. ${}_7P_1 =$

A. 6! B. 7! C. 6 D. 7 E. 1

9-24. ${}_{7}P_{0} =$

A. 6! B. 7! C. 6 D. 7 E. 1

9-25. ${}_{7}C_{7} =$

A. 6! B. 7! C. 6 D. 7 E. 1

Test form 9B Date _________ Name ___________________________
Course ______ Sec ______ Student Number ________________

Chapter 9

9-1. 9!

A. 362880 B. 9 C. 729

D. 81 E. 362881

9-2. ${}_8P_2 =$

A. 56 B. 8 C. 6 D. 336 E. 4

9-3. ${}_8C_7 =$

A. 7 B. 56 C. 28 D. 8 E. 2

9-4. A person has 9 colors of paint and 8 colors of floor tile to choose from. If one of each are to be selected, how many color schemes are possible for a room?

A. 72 B. 73 C. 74 D. 75 E. 76

9-5. A person must select one type of drink and one snack at the movies. If there are 5 different drinks and 8 different snacks available, how many possibilities are there?

A. 38 B. 39 C. 40 D. 41 E. 42

9-6. If there are 30 students in a class and each is to select a different special assignment from 30 that are available, in how many ways can the selection be made?

A. 30! B. ${}_{30}P_1$ C. ${}_{30}C_{30}$

D. ${}_{30}C_0$ E. 30

9-7. In how many ways can the top three of 25 teams be awarded prizes in an academic competition?

A. 25! B. ${}_{25}P_3$ C. ${}_{25}C_3$

D. ${}_{25}C_0$ E. $\frac{25}{3}$

9-8. In how many ways can a student select 10 of 12 questions on a test?

A. 10^{12} B. 12^{10} C. ${}_{12}P_{10}$ D. ${}_{12}C_{10}$ E. $\frac{10!}{12!}$

9-9. Of 12 students 10 are to be given identical awards. In how many ways can this be done?

A. 10^{12} B. 12^{10} C. ${}_{12}P_{10}$ D. ${}_{12}C_{10}$ E. $\frac{10!}{12!}$

9-10. A 6-sided die is rolled. What is the probability the side up will be 3?

A. 3 B. $\frac{1}{3}$ C. 6 D. $\frac{1}{6}$ E. 0

9-11. A fair coin is tossed 9 times. What is the probability of getting 4 heads?

A. $\frac{1}{2^9}$ B. $\frac{{}_9P_4}{2^9}$ C. $\frac{{}_9C_4}{2^9}$

D. $\frac{1}{2}$ E. $\frac{4}{9}$

9-12. If one card is selected from a deck of 52 cards, what is the probability it is a club?

A. $\frac{1}{52}$ B. $\frac{1}{4}$ C. $\frac{1}{13}$ D. $\frac{1}{48}$ E. $\frac{1}{12}$

9-13. Consider the experiment of tossing two dice. Determine the probability that the sum that comes up is 5.

A. $\frac{5}{36}$ B. $\frac{1}{36}$ C. $\frac{2}{36}$ D. $\frac{4}{36}$ E. $\frac{35}{36}$

9-14. Two cards are selected from a deck of 52 cards. What is the probability one is a 10 and the other is an ace?

A. $\frac{4}{52} \cdot \frac{4}{51}$ B. $\frac{1}{52} \cdot \frac{1}{51}$

C. $\frac{4}{52} \cdot \frac{4}{52}$ D. $\frac{4}{52} \cdot \frac{3}{51}$

E. $2 \cdot \frac{4}{52} \cdot \frac{4}{51}$

9-15. A card is selected from a deck of 52 cards. What is the probability of getting a 6 or a diamond?

A. $\frac{1}{52}$ B. $\frac{1}{13}$ C. $\frac{1}{4}$ D. $\frac{4}{13}$ E. $\frac{1}{3}$

9-16. A card is selected from a deck of 52 cards. What is the probability of getting a 6 and a club?

A. $\frac{1}{52}$ B. $\frac{1}{13}$ C. $\frac{1}{4}$ D. $\frac{4}{13}$ E. $\frac{1}{3}$

9-17. A card is selected from a deck of 52 cards. What is the probability of getting a jack or a 9?

A. $\frac{1}{52}$ B. $\frac{1}{4}$ C. $\frac{2}{13}$ D. $\frac{1}{13}$ E. $\frac{1}{3}$

9-18. Suppose there are 8 freshmen, 11 sophomores, and 6 juniors in a group. One student is selected at random. What is the probability a junior is selected?

A. $\frac{8}{25}$ B. $\frac{11}{25}$ C. $\frac{6}{25}$ D. $\frac{8}{26}$ E. $\frac{11}{26}$

9-19. What is the probability of being dealt a 5-card hand with 4 kings?

A. $\frac{4}{52}$ B. $\frac{{}_{48}C_1}{{}_{52}C_5}$ C. $13\frac{{}_{48}C_1}{{}_{52}C_5}$

D. $\frac{{}_{52}C_4}{{}_{52}C_5}$ E. $13\frac{{}_{52}C_4}{{}_{52}C_5}$

9-20. A box contains 21 good radios and 5 defective radios. What is the probability of getting exactly one defective radio if 3 radios are selected?

A. $\dfrac{{}_5C_1}{{}_{21}C_3}$ B. $\dfrac{{}_{21}C_3}{{}_{26}C_3}$ C. $\dfrac{{}_{21}C_2\ {}_5C_1}{{}_{26}C_3}$

D. $\dfrac{{}_5C_1}{{}_{26}C_2}$ E. $\dfrac{{}_{21}C_1}{{}_{26}C_2}$

9-21. A box contains 21 good radios and 5 defective radios. What is the probability of getting at most one defective radio if 3 radios are selected?

A. $\dfrac{{}_{21}C_2\ {}_5C_1}{{}_{26}C_3}+\dfrac{{}_{21}C_3}{{}_{26}C_3}$ B. $\dfrac{{}_{21}C_2\ {}_5C_1}{{}_{26}C_3}$

C. $\dfrac{{}_5C_1}{{}_{21}C_3}$ D. $\dfrac{{}_{21}C_1}{{}_{26}C_2}$

E. $\dfrac{{}_5C_1}{{}_{26}C_2}$

9-22. A box contains 21 good radios and 5 defective radios. What is the probability of getting at least two defective radios if 3 radios are selected?

A. $1-\dfrac{{}_{21}C_2\ {}_5C_1}{{}_{26}C_3}-\dfrac{{}_{21}C_3}{{}_{26}C_3}$ B. $1-\dfrac{{}_{21}C_2\ {}_5C_1}{{}_{26}C_3}$

C. $\dfrac{{}_5C_1}{{}_{21}C_3}$ D. $\dfrac{{}_{21}C_1}{{}_{26}C_2}$

E. $\dfrac{{}_5C_1}{{}_{26}C_2}$

9-23. ${}_8P_1 =$

A. 8! B. 7! C. 8 D. 7 E. 1

9-24. ${}_8P_0 =$

A. 8! B. 7! C. 8 D. 7 E. 1

9-25. ${}_8C_8 =$

A. 8! B. 7! C. 8 D. 7 E. 1

Test form 9C Date _________ Name ___________________________
Course ______ Sec ______ Student Number ________________

Chapter 9

9-1. 4!

A. 4 B. 16 C. 24 D. 25 E. 64

9-2. ${}_{10}P_2 =$

A. 10 B. 8 C. 720 D. 90 E. 5

9-3. ${}_{9}C_8 =$

A. 8 B. 9 C. 72 D. 36 E. 2

9-4. A person has 10 colors of paint and 8 colors of floor tile to choose from. If one of each are to be selected, how many color schemes are possible for a room?

A. 79 B. 80 C. 81 D. 82 E. 83

9-5. A person must select one type of drink and one snack at the movies. If there are 2 different drinks and 7 different snacks available, how many possibilities are there?

A. 12 B. 13 C. 14 D. 15 E. 16

9-6. If there are 20 students in a class and each is to select a different special assignment from 20 that are available, in how many ways can the selection be made?

A. 20! B. ${}_{20}P_1$ C. ${}_{20}C_{20}$

D. ${}_{20}C_0$ E. 20

9-7. In how many ways can the top three of 28 teams be awarded prizes in an academic competition?

A. 28! B. ${}_{28}P_3$ C. ${}_{28}C_3$

D. ${}_{28}C_0$ E. $\frac{28}{3}$

9-8. In how many ways can a student select 5 of 10 questions on a test?

A. 5^{10} B. 10^5 C. ${}_{10}P_5$ D. ${}_{10}C_5$ E. $\frac{5!}{10!}$

9-9. Of 10 students 5 are to be given identical awards. In how many ways can this be done?

A. 5^{10} B. 10^5 C. ${}_{10}P_5$ D. ${}_{10}C_5$ E. $\frac{5!}{10!}$

9-10. A 6-sided die is rolled. What is the probability the side up will be 4?

A. 4 B. $\frac{1}{4}$ C. 6 D. $\frac{1}{6}$ E. 0

9-11. A fair coin is tossed 10 times. What is the probability of getting 7 heads?

A. $\frac{1}{2^{10}}$ B. $\frac{{}_{10}P_7}{2^{10}}$ C. $\frac{{}_{10}C_7}{2^{10}}$

D. $\frac{1}{2}$ E. $\frac{7}{10}$

9-12. If one card is selected from a deck of 52 cards, what is the probability it is a 5?

A. $\frac{1}{52}$ B. $\frac{1}{4}$ C. $\frac{1}{13}$ D. $\frac{1}{48}$ E. $\frac{1}{12}$

9-13. Consider the experiment of tossing two dice. Determine the probability that the sum that comes up is 10.

A. $\frac{10}{36}$ B. $\frac{3}{36}$ C. $\frac{1}{36}$ D. $\frac{2}{36}$ E. $\frac{35}{36}$

9-14. Two cards are selected from a deck of 52 cards. What is the probability one is a 6 and the other is an ace?

A. $\frac{4}{52}\cdot\frac{4}{51}$ B. $\frac{1}{52}\cdot\frac{1}{51}$

C. $\frac{4}{52}\cdot\frac{4}{52}$ D. $\frac{4}{52}\cdot\frac{3}{51}$

E. $2\cdot\frac{4}{52}\cdot\frac{4}{51}$

9-15. A card is selected from a deck of 52 cards. What is the probability of getting a jack or a diamond?

A. $\frac{1}{52}$ B. $\frac{1}{13}$ C. $\frac{1}{4}$ D. $\frac{4}{13}$ E. $\frac{1}{3}$

9-16. A card is selected from a deck of 52 cards. What is the probability of getting a 10 and a heart?

A. $\frac{1}{4}$ B. $\frac{1}{13}$ C. $\frac{1}{52}$ D. $\frac{4}{13}$ E. $\frac{1}{3}$

9-17. A card is selected from a deck of 52 cards. What is the probability of getting a 7 or a 6?

A. $\frac{1}{52}$ B. $\frac{1}{13}$ C. $\frac{1}{3}$ D. $\frac{2}{13}$ E. $\frac{1}{4}$

9-18. Suppose there are 11 freshmen, 6 sophomores, and 8 juniors in a group. One student is selected at random. What is the probability a junior is selected?

A. $\frac{11}{25}$ B. $\frac{6}{25}$ C. $\frac{8}{25}$ D. $\frac{11}{27}$ E. $\frac{6}{27}$

9-19. What is the probability of being dealt a 5-card hand with 4 aces?

A. $\frac{4}{52}$ B. $\frac{{}_{48}C_1}{{}_{52}C_5}$ C. $13\frac{{}_{48}C_1}{{}_{52}C_5}$

D. $\frac{{}_{52}C_4}{{}_{52}C_5}$ E. $13\frac{{}_{52}C_4}{{}_{52}C_5}$

9-20. A box contains 25 good radios and 5 defective radios. What is the probability of getting exactly one defective radio if 3 radios are selected?

A. $\frac{{}_{5}C_1}{{}_{30}C_2}$ B. $\frac{{}_{25}C_3}{{}_{30}C_3}$ C. $\frac{{}_{5}C_1}{{}_{25}C_3}$

D. $\frac{{}_{25}C_1}{{}_{30}C_2}$ E. $\frac{{}_{25}C_2\ {}_{5}C_1}{{}_{30}C_3}$

9-21. A box contains 25 good radios and 5 defective radios. What is the probability of getting at most one defective radio if 3 radios are selected?

A. $\frac{{}_{25}C_2\ {}_{5}C_1}{{}_{30}C_3}$ B. $\frac{{}_{25}C_1}{{}_{30}C_2}$

C. $\frac{{}_{5}C_1}{{}_{25}C_3}$ D. $\frac{{}_{25}C_2\ {}_{5}C_1}{{}_{30}C_3}+\frac{{}_{25}C_3}{{}_{30}C_3}$

E. $\frac{{}_{5}C_1}{{}_{25}C_2}$

9-22. A box contains 25 good radios and 5 defective radios. What is the probability of getting at least one defective radio if 3 radios are selected?

A. $1-\frac{{}_{25}C_2\ {}_5C_1}{{}_{30}C_3}-\frac{{}_{25}C_3}{{}_{30}C_3}$

B. $1-\frac{{}_{25}C_2\ {}_5C_1}{{}_{30}C_3}$

C. $\frac{{}_5C_1}{{}_{25}C_3}$

D. $\frac{{}_{25}C_1}{{}_{30}C_2}$

E. $\frac{{}_5C_1}{{}_{30}C_2}$

9-23. ${}_{10}P_1 =$

A. 10! B. 9! C. 1 D. 9 E. 10

9-24. ${}_{10}P_0 =$

A. 10! B. 9! C. 1 D. 9 E. 10

9-25. ${}_{10}C_{10} =$

A. 10! B. 9! C. 1 D. 9 E. 10

Test form 9D   Date ________
Course ______ Sec ______

Name ________________________
Student Number _______________

Chapter 9

9-1. 12!

9-2. ${}_7P_3 =$

9-3. ${}_{15}C_{14} =$

9-4. A person has 12 colors of paint and 4 colors of floor tile to choose from. If one of each are to be selected, how many color schemes are possible for a room?

9-5. A person must select one type of drink and one snack at the movies. If there are 4 different drinks and 9 different snacks available, how many possibilities are there?

9-6. If there are 8 students in a class and each is to select a different special assignment from 8 that are available, in how many ways can the selection be made?

9-7. In how many ways can the top three of 10 teams be awarded prizes in an academic competition?

9-8. In how many ways can a student select 5 of 8 questions on a test?

9-9. Of 8 students 5 are to be given identical awards. In how many ways can this be done?

9-10. A 4-sided die with sides labeled 1 through 4 is rolled. What is the probability the side up will be 1?

9-11. A fair coin is tossed 12 times. What is the probability of getting 4 heads?

9-12. If one card is selected from a deck of 52 cards, what is the probability it is a 10?

9-13. Consider the experiment of tossing two dice. Determine the probability that the sum that comes up is 9.

9-14. Two cards are selected from a deck of 52 cards. What is the probability one is a 4 and the other is an 8?

9-15. A card is selected from a deck of 52 cards. What is the probability of getting a 10 or a club?

9-16. A card is selected from a deck of 52 cards. What is the probability of getting a 7 and a club?

9-17. A card is selected from a deck of 52 cards. What is the probability of getting a king or a jack?

9-18. Suppose there are 9 freshmen, 4 sophomores, and 12 juniors in a group. One student is selected at random. What is the probability a junior is selected?

9-19. What is the probability of being dealt a 5-card hand with 4 6s?

9-20. A box contains 40 good radios and 5 defective radios. What is the probability of getting exactly one defective radio if 3 radios are selected?

9-21. A box contains 40 good radios and 5 defective radios. What is the probability of getting at most one defective radio if 3 radios are selected?

9-22. A box contains 40 good radios and 5 defective radios. What is the probability of getting at least two defective radios if 3 radios are selected?

9-23. ${}_6P_1 =$

9-24. ${}_6P_0 =$

9-25. ${}_6C_6 =$

Test form 9E　　　　Date ________　　　　Name ______________________
Course ______ Sec ______　　　　Student Number ______________

Chapter 9

9-1. 5!

9-2. ${}_6P_4 =$

9-3. ${}_{12}C_{11} =$

9-4. A person has 10 colors of paint and 8 colors of floor tile to choose from. If one of each are to be selected, how many color schemes are possible for a room?

9-5. A person must select one type of drink and one snack at the movies. If there are 4 different drinks and 7 different snacks available, how many possibilities are there?

9-6. If there are 9 students in a class and each is to select a different special assignment from 9 that are available, in how many ways can the selection be made?

9-7. In how many ways can the top three of 9 teams be awarded prizes in an academic competition?

9-8. In how many ways can a student select 6 of 9 questions on a test?

9-9. Of 9 students 6 are to be given identical awards. In how many ways can this be done?

9-10. A 4-sided die with sides labeled 1 through 4 is rolled. What is the probability the side up will be 2?

9-11. A fair coin is tossed 16 times. What is the probability of getting 5 heads?

9-12. If one card is selected from a deck of 52 cards, what is the probability it is a king?

9-13. Consider the experiment of tossing two dice. Determine the probability that the sum that comes up is 3.

9-14. Two cards are selected from a deck of 52 cards. What is the probability one is a 6 and the other is an ace?

9-15. A card is selected from a deck of 52 cards. What is the probability of getting a 6 or a black card?

9-16. A card is selected from a deck of 52 cards. What is the probability of getting a 9 and a club?

9-17. A card is selected from a deck of 52 cards. What is the probability of getting a 3 or a 4?

9-18. Suppose there are 4 freshmen, 6 sophomores, and 10 juniors in a group. One student is selected at random. What is the probability a junior is selected?

9-19. What is the probability of being dealt a 5-card hand with 4 9s?

9-20. A box contains 30 good radios and 6 defective radios. What is the probability of getting exactly one defective radio if 3 radios are selected?

9-21. A box contains 30 good radios and 6 defective radios. What is the probability of getting at most one defective radio if 3 radios are selected?

9-22. A box contains 30 good radios and 6 defective radios. What is the probability of getting at least two defective radios if 3 radios are selected?

9-23. ${}_{10}P_1 =$

9-24. ${}_{10}P_0 =$

9-25. ${}_{10}C_{10} =$

Test form 9F Date ________ Name ________________________
Course ______ Sec ______ Student Number ________________

Chapter 9

9-1. 8!

9-2. ${}_{9}P_{3} =$

9-3. ${}_{7}C_{6} =$

9-4. A person has 6 colors of paint and 8 colors of floor tile to choose from. If one of each are to be selected, how many color schemes are possible for a room?

9-5. A person must select one type of drink and one snack at the movies. If there are 5 different drinks and 9 different snacks available, how many possibilities are there?

9-6. If there are 11 students in a class and each is to select a different special assignment from 11 that are available, in how many ways can the selection be made?

9-7. In how many ways can the top three of 8 teams be awarded prizes in an academic competition?

9-8. In how many ways can a student select 3 of 8 questions on a test?

9-9. Of 8 students 3 are to be given identical awards. In how many ways can this be done?

9-10. A 4-sided die with sides labeled 1 through 4 is rolled. What is the probability the side up will be 3?

9-11. A fair coin is tossed 11 times. What is the probability of getting 4 heads?

9-12. If one card is selected from a deck of 52 cards, what is the probability it is a spade?

9-13. Consider the experiment of tossing two dice. Determine the probability that the sum that comes up is 11.

9-14. Two cards are selected from a deck of 52 cards. What is the probability one is a jack and the other is an 8?

9-15. A card is selected from a deck of 52 cards. What is the probability of getting a 9 or a red card?

9-16. A card is selected from a deck of 52 cards. What is the probability of getting a jack and a spade?

9-17. A card is selected from a deck of 52 cards. What is the probability of getting a queen or a king?

9-18. Suppose there are 7 freshmen, 4 sophomores, and 8 juniors in a group. One student is selected at random. What is the probability a junior is selected?

9-19. What is the probability of being dealt a 5-card hand with 4 4s?

9-20. A box contains 45 good radios and 5 defective radios. What is the probability of getting exactly one defective radio if 3 radios are selected?

9-21. A box contains 45 good radios and 5 defective radios. What is the probability of getting at most one defective radio if 3 radios are selected?

9-22. A box contains 45 good radios and 5 defective radios. What is the probability of getting at least two defective radios if 3 radios are selected?

9-23. ${}_{13}P_1 =$

9-24. ${}_{13}P_0 =$

9-25. ${}_{13}C_{13} =$

Test form Final A Date ________ Name ______________________

Course ______ Sec ______ Student Number ______________

Final Exam

1. Find the x- and y-intercepts of the graph of $2x + 4y = 24$ by letting y and then x be zero.

   A. x-intercept (12, 0), y-intercept (0, 6)

   B. x-intercept (6, 0), y-intercept (0, 12)

   C. x-intercept (–12, 0), y-intercept (0, –6)

   D. x-intercept (–6, 0), y-intercept (0, –12)

   E. x-intercept (2, 0), y-intercept (0, 4)

2. Use the point-slope form, $y - y_1 = m(x - x_1)$, with $m = \frac{y_2 - y_1}{x_2 - x_1}$ to find the equation of the line through the points (12, 16) and (20, 32).

   A. $y-16=-2(x-12)$ B. $y-16=\frac{1}{2}(x-12)$

   C. $y-16=2(x-12)$ D. $y-12=2(x-16)$

   E. $y-12=-2(x-16)$

3. Find the vertex of $y=3x^2+12x-7$.

   A. (3, 12) B. (3, 7) C. (–3, –12)

   D. (–2, –12) E. (–2, –19)

4. Find the points of intersection of $y = 2x^2 - 4$ and $y = -3x^2$.

A. (–0.8944, –2.4) and (0.8944, –2.4)

B. (–0.8901, –2.4) and (0.8901, –2.4)

C. (–0.8956, –2.4) and (0.8956, –2.4)

D. (–0.8972, –2.4) and (0.8972, –2.4)

E. (–0.8991, –2.4) and (0.8991, –2.4)

5. Determine the equation of the circle having center (–6, 10) and radius 4.

A. $(x+6)^2 + (y+10)^2 = 4$ B. $(x-6)^2 + (y+10)^2 = 4$

C. $(x+6)^2 + (y-10)^2 = 4$ D. $(x+6)^2 + (y-10)^2 = 16$

E. $(x-6)^2 + (y+10)^2 = 16$

6. Solve the quadratic equation. $x^2 + 13x + 36 = 0$

A. $x = 4, 9$ B. $x = -4, -9$ C. $x = -3, -12$

D. $x = 3, 12$ E. $x = -18, -2$

7. Use the equation $s = -16t^2 + vt$ that gives the height s feet at time t seconds of an object fired upward with a velocity v feet per second. If the object is propelled at 240 feet per second, when will it come back to the ground?

A. 14 seconds B. 15 seconds C. 16 seconds

D. 17 seconds E. 18 seconds

8. Solve the equation $\sqrt{x-4} = 3$.

A. 13 B. 14 C. 15

D. 16 E. No solution

9. Solve the linear inequality. Write the solution in interval notation.
$9x - 7 > 3x + 20$

A. $(-\infty, -4.5)$ B. $(-\infty, 4.5)$ C. $(4.5, \infty)$

D. $(-4.52, \infty)$ E. $(-4.5, 4.5)$

10. Solve the inequality. Write the solution in interval notation. $|x| < 21$

A. $(-\infty, -21)$ B. $(-\infty, 21)$ C. $(-21, 21)$

D. $(-21, \infty)$ E. $(21, \infty)$

11. Determine the domain and range of $f(x) = x^2 + 6x - 2$.

A. Domain: $(-\infty, -3)$; Range: $[-11, \infty)$

B. Domain: $(-\infty, \infty)$; Range: $[-11, \infty)$

C. Domain: $(-\infty, \infty)$; Range: $(-11, \infty)$

D. Domain: $(-\infty, -3)$; Range: $(-11, \infty)$

E. Domain: $(-3, -11)$; Range: $(-\infty, \infty)$

12. Determine the domain of $f(x) = \dfrac{(14x + 3)(x - 2)}{x - 2}$.

A. All real numbers

B. All real numbers except -2

C. All real numbers except $(-2, 2)$

D. All real numbers except 2

E. 2

13. $f(x) = x^2$ and $g(x) = x^2 - 2x$. Find $(g \circ f)(x)$.

A. $2x^2 - 2x$ B. $-2x$ C. $x^4 - 2x^2$

D. $(x^2 - 2x)^2$ E. $\sqrt{x^2 - 2x}$

14. Of $f(x), g(x)$ and $h(x)$, list all one-to-one functions.

$f(x)$

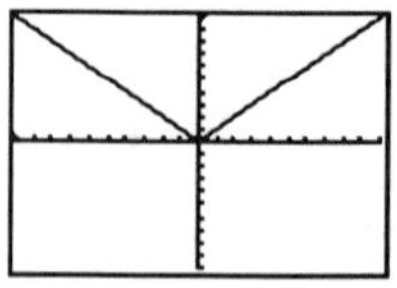

$g(x)$

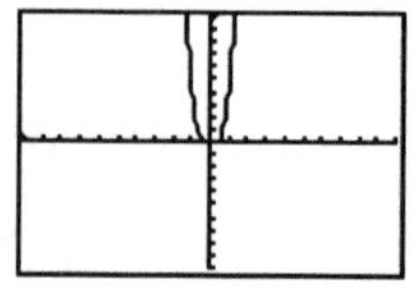

$h(x)$

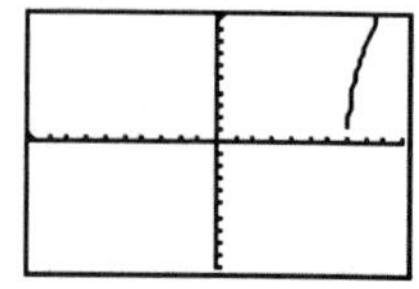

A. $f(x), g(x)$, and $h(x)$

B. $f(x)$

C. $h(x)$

D. $g(x)$

E. $g(x)$, and $h(x)$

15. Find the inverse function for $f(x) = 2x^9 + 7$.

A. $f^{-1}(x) = \dfrac{\sqrt[9]{x-7}}{2}$

B. $f^{-1}(x) = \dfrac{\sqrt[9]{x}-7}{2}$

C. $f^{-1}(x) = \sqrt[9]{\dfrac{x-7}{2}}$

D. $f^{-1}(x) = \dfrac{\sqrt[9]{x}}{2} - 7$

E. Does not exist.

16. Compute the product $(4+6i)(3+5i)$.

A. $18 + 38i$

B. $-18 + 38i$

C. $42 + 38i$

D. $-42 + 38i$

E. $12 + 30i$

17. Perform the division $\dfrac{7+2i}{1-i}$.

A. $\dfrac{5+9i}{2}$

B. $\dfrac{-5+9i}{2}$

C. $\dfrac{-5-9i}{2}$

D. $\dfrac{5-9i}{2}$

E. $-2i$

18. Perform the additions and subtractions. Write the answer in the standard form $a+bi$. $(3+8i)-(7+3i)+(9-2i)$

A. $2+5i$ B. $9+3i$ C. $1+4i$

D. $5+3i$ E. $7+i$

19. Find the real zeros of $f(x)=x^4-6x^3-x^2+54x-72$.

A. 2, 4 B. 2, 3, 4

C. −3, 2, 3, 4 D. −3, 3

E. −4, −3, −2, 3

20. Solve using algebraic methods. $6^{x-4}=6^{3x}$

A. −4 B. 4 C. 2 D. −2 E. −1

21. Solve the equation (if possible) to four decimal places. $9^{2x+4}=7$

A. −1.5592 B. −1.5572 C. −1.5564

D. −1.5551 E. −1.5543

22. Compute $\log_3 6561$.

A. 5 B. 6 C. 7 D. 8 E. 9

23. Use the properties of logarithms to solve the equation for x. $\log(x-5)=2$

A. 90 B. 95 C. 100 D. 105 E. No solution

24. Solve the equation for x, to four decimal places, by writing it in logarithmic form, then use the calculator. $7=e^{x/3}$

A. 5.8341 B. 5.8377 C. 5.8392

D. 5.8402 E. 5.8421

25. A person has 30 coins with a value of \$4.70. If the coins are all quarters and nickels, how many of each does the person have?

A. 16 quarters, 14 nickels
B. 16 quarters, 15 nickels
C. 15 quarters, 15 nickels
D. 14 quarters, 16 nickels
E. 12 quarters, 18 nickels

26. Starting with $\begin{bmatrix} 1 & 1 & 2 & 4 \\ 5 & 6 & 8 & 2 \\ 7 & 1 & 3 & 6 \end{bmatrix}$, perform the transformation $-3R_1 + R_2$.

A. $\begin{bmatrix} 1 & 1 & 2 & 4 \\ 2 & 3 & 2 & -10 \\ 7 & 1 & 3 & 6 \end{bmatrix}$

B. $\begin{bmatrix} 0 & 0 & 0 & 0 \\ 2 & 3 & 2 & -10 \\ 7 & 1 & 3 & 6 \end{bmatrix}$

C. $\begin{bmatrix} -3 & -3 & -6 & -12 \\ 5 & 6 & 8 & 2 \\ 7 & 1 & 3 & 6 \end{bmatrix}$

D. $\begin{bmatrix} -3 & -3 & -6 & -12 \\ 2 & 3 & 2 & -10 \\ 7 & 1 & 3 & 6 \end{bmatrix}$

E. $\begin{bmatrix} 2 & 3 & 2 & -10 \\ 7 & 1 & 3 & 6 \end{bmatrix}$

27. Solve (if possible) the system of three equations in three variables, using the method of Gauss-Jordan elimination.

$$5x + 2y + 3z = 18$$
$$x - y + 3z = 7$$
$$4x + y - z = 7$$

A. (2, 1, 2)
B. (1, 3, 2)
C. (3, 1, 0)
D. (0, 4, 4)
E. No solution

28. A company makes two different products. Product A requires 20 hours of assembly and 5 pounds of a certain raw material, while product B requires 10 hours of assembly and 10 pounds of the same raw material. There are 1,200 hours of assembly and 1,000 pounds of the raw material available. If the profit from each unit of product A is \$350 and from each unit of product B is \$400, what is the maximum profit the company can make?

A. \$35,000 B. \$12,400 C. \$45,000

D. \$24,000 E. \$42,000

29. Solve the matrix equation $\begin{bmatrix} x & x+z \\ y & y+z \end{bmatrix} = \begin{bmatrix} 5 & 8 \\ 6 & 9 \end{bmatrix}$.

A. (5, 6, 3) B. (5, 6, 8) C. (5, 6, 2)

D. (5, 6, 9) E. (5, 6, 0)

30. Perform the multiplication $\begin{bmatrix} 9 & 3 \\ 5 & 7 \end{bmatrix}\begin{bmatrix} 7 & 6 \\ 4 & 8 \end{bmatrix}$, if possible.

A. $\begin{bmatrix} 63 & 18 \\ 20 & 56 \end{bmatrix}$ B. $\begin{bmatrix} 75 & 78 \\ 63 & 86 \end{bmatrix}$ C. $\begin{bmatrix} 78 & 75 \\ 86 & 63 \end{bmatrix}$

D. $\begin{bmatrix} 86 & 78 \\ 63 & 75 \end{bmatrix}$ E. Not possible

31. Compute the determinant $\begin{vmatrix} 5 & 9 \\ 7 & 9 \end{vmatrix}$.

A. 18 B. −18 C. 17 D. −17 E. 0

32. Evaluate $\sum_{k=1}^{5}(k^2+1)$.

A. 60 B. 61 C. 62 D. 63 E. 64

33. The first term $a_1 = 7$ and the common difference $d = -2$ of an arithmetic sequence are given. Find the first four terms.

A. 7, 5, 3, 1 B. 7, −9, 11, −13

C. 7, −5, 3, −1 D. 7, −5, 3, 1

E. 5, 3, 1, −1

34. The first term of an arithmetic sequence is $a_1 = 7$ and the difference is $d = 4$. Find the 5th term.

A. 25 B. 21 C. 23 D. 27 E. 31

35. The first term and the common ratio of a geometric sequence are $a_1 = 5$ and $r = 8$. Find the 8th term.

A. $5(8)^7$ B. $5(8)^8$ C. $5(8)^9$ D. $(40)^7$ E. $(40)^7$

36. Expand $(a+2b)^5$ using Pascal's triangle.

A. $a^5 + 2a^4b + 4a^3b^2 + 8a^2b^3 + 16ab^4 + 32b^5$

B. $a^5 + 5a^4b + 10a^3b^2 + 10a^2b^3 + 5ab^4 + b^5$

C. $a^5 + 10a^4b + 40a^3b^2 + 80a^2b^3 + 80ab^4 + 32b^5$

D. $a^5 - 10a^4b + 40a^3b^2 - 80a^2b^3 + 80ab^4 - 32b^5$

E. $a^5 - 5a^4b + 10a^3b^2 - 10a^2b^3 + 5ab^4 - b^5$

37. ${}_{20}P_3 =$

A. 6,800 B. 6,900 C. 7,220 D. 7,200 E. 6,840

38. In how many ways can the top three of 60 teams be awarded prizes in an academic competition?

A. 60! B. ${}_{60}P_3$ C. ${}_{60}C_3$

D. ${}_{60}C_0$ E. 20

39. If one card is selected from a deck of 52 cards, what is the probability it is the king of spades?

A. $\frac{1}{13}$ B. $\frac{1}{4}$ C. $\frac{1}{12}$ D. $\frac{1}{52}$ E. 1

40. A box contains 86 good radios and 4 defective radios. What is the probability of getting at most one defective radio if 3 radios are selected?

A. $\frac{{}_{86}C_2\;{}_4C_1}{{}_{90}C_3} + \frac{{}_{86}C_3}{{}_{90}C_3}$

B. $\frac{{}_{86}C_2\;{}_4C_1}{{}_{90}C_3}$

C. $\frac{{}_4C_1}{{}_{90}C_3}$

D. $\frac{86}{90}$

E. $\frac{1}{4}$

Test form Final B   Date ________   Name ____________________________

Course ______ Sec ______   Student Number ________________

Final Exam

1. Find the x- and y-intercepts of the graph of $3x+4y=24$ by letting y and then x be zero.

   A. x-intercept (8, 0), y-intercept (0, 6)

   B. x-intercept (6, 0), y-intercept (0, 8)

   C. x-intercept (–8, 0), y-intercept (0, 8)

   D. x-intercept (–6, 0), y-intercept (0, –8)

   E. x-intercept (3, 0), y-intercept (0, 4)

2. Use the point-slope form, $y-y_1=m(x-x_1)$, with $m=\dfrac{y_2-y_1}{x_2-x_1}$ to find the equation of the line through the points (15, 30) and (30, 60).

   A. $y-15=2(x-30)$ B. $y-15=-2(x-30)$

   C. $y-30=-2(x-15)$ D. $y-30=2(x-15)$

   E. $y-30=\frac{1}{2}(x-15)$

3. Find the vertex of $y=4x^2+12x-5$

   A. (4, 5)   B. (–4, –12)   C. (4, 12)

   D. (–1.5, –14)   E. (–1.5, –11)

4. Find the points of intersection of $y = 2x^2 - 6$ and $y = -3x^2$.

A. (−1.0954, −3.6) and (1.0954, −3.6)

B. (−1.0984, −3.6) and (1.0984, −3.6)

C. (−1.0945, −3.6) and (1.0945, −3.6)

D. (−1.0992, −3.6) and (1.0992, −3.6)

E. (−1.0932, −3.6) and (1.0932, −3.6)

5. Determine the equation of the circle having center (−7, 12) and radius 5.

A. $(x+7)^2 + (y+12)^2 = 5$ B. $(x-7)^2 + (y+12)^2 = 5$

C. $(x+7)^2 + (y-12)^2 = 5$ D. $(x-7)^2 + (y+12)^2 = 25$

E. $(x+7)^2 + (y-12)^2 = 25$

6. Solve the quadratic equation. $x^2 - 13x + 36 = 0$

A. $x = 4, 9$ B. $x = -4, -9$ C. $x = -3, -12$

D. $x = 3, 12$ E. $x = -18, -2$

7. Use the equation $s = -16t^2 + vt$ that gives the height s feet at time t seconds of an object fired upward with a velocity v feet per second. If the object is propelled at 256 feet per second, when will it come back to the ground?

A. 14 seconds B. 15 seconds C. 16 seconds

D. 17 seconds E. 18 seconds

8. Solve the equation $\sqrt{x+3} = 8$.

A. 57 B. 58 C. 59

D. 61 E. No solution

9. Solve the linear inequality. Write the solution in interval notation.
$6x - 15 > 3x + 30$

A. $(-\infty, -15)$ B. $(-\infty, 15)$ C. $(-15, \infty)$

D. $(15, \infty)$ E. $(-15, 15)$

10. Solve the inequality. Write the solution in interval notation. $|x| < 29$

A. $(-\infty, -29)$ B. $(-\infty, 29)$ C. $(-29, 29)$

D. $(-29, \infty)$ E. $(29, \infty)$

11. Determine the domain and range of $f(x) = x^2 + 8x - 2$.

A. Domain: $(-\infty, \infty)$; Range: $[-18, \infty)$

B. Domain: $(-\infty, -4)$; Range: $[-18, \infty)$

C. Domain: $(-\infty, \infty)$; Range: $(-18, \infty)$

D. Domain: $(-\infty, -4)$; Range: $(-18, \infty)$

E. Domain: $(-4, -18)$; Range: $(-\infty, \infty)$

12. Determine the domain of $f(x) = \dfrac{(19x+7)(x-2)}{x-2}$.

A. All real numbers

B. All real numbers except -2

C. All real numbers except $(-2, 2)$

D. All real numbers except 2

E. 2

13. $f(x) = x^3$ and $g(x) = x^3 - 4x$. Find $(g \circ f)(x)$.

A. $2x^3 - 4x$ B. $-4x$ C. $x^9 - 4x^3$

D. $(x^3 - 4x)^3$ E. $\sqrt[3]{x^3 - 4x}$

14. Of $f(x), g(x)$ and $h(x)$, list all one-to-one functions.

$f(x)$ 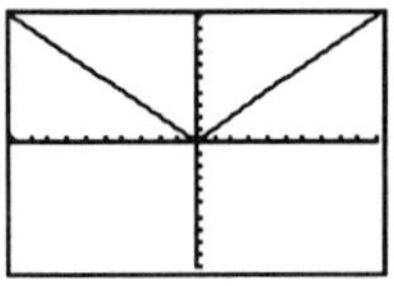

$g(x)$ 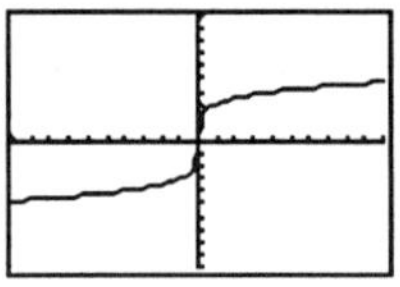

$h(x)$ 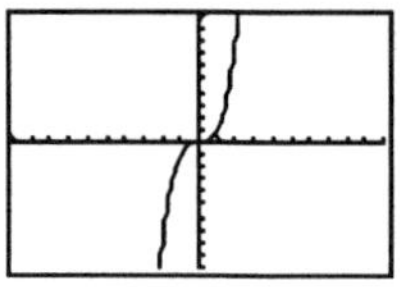

A. $f(x), g(x)$, and $h(x)$ B. $f(x)$

C. $h(x)$ D. $g(x)$

E. $g(x)$, and $h(x)$

15. Find the inverse function for $f(x) = 3x^9 + 8$.

A. $f^{-1}(x) = \dfrac{\sqrt[9]{x-8}}{3}$ B. $f^{-1}(x) = \dfrac{\sqrt[9]{x}-8}{3}$

C. $f^{-1}(x) = \sqrt[9]{\dfrac{x-8}{3}}$ D. $f^{-1}(x) = \dfrac{\sqrt[9]{x}}{3} - 8$

E. Does not exist.

16. Compute the product $(5+4i)(3+7i)$.

A. $13+47i$ B. $43+47i$ C. $-43+47i$

D. $-13+47i$ E. $15+28i$

17. Perform the division $\dfrac{11+2i}{1-i}$.

A. $\dfrac{9-13i}{2}$ B. $\dfrac{-9-13i}{2}$ C. $\dfrac{9+13i}{2}$

D. $\dfrac{-9+13i}{2}$ E. $-2i$

18. Perform the additions and subtractions. Write the answer in the standard form $a + bi$. $(9+7i)-(5+6i)+(10-5i)$

A. $10+3i$ B. $14-4i$ C. $12-6i$

D. $9+7i$ E. $9+3i$

19. Find the real zeros of $f(x) = x^4 - 7x^3 + x^2 + 63x - 90$.

A. 2, 5 B. 2, 3, 5

C. −3, 2, 3, 5 D. −3, 3

E. −5, −3, −2, 3

20. Solve using algebraic methods. $7^{x-8} = 7^{3x}$

A. −4 B. 4 C. −8 D. 8 E. −2

21. Solve the equation (if possible) to four decimal places. $7^{2x+6} = 3$

A. −2.7128 B. −2.7136 C. −2.7151

D. −2.7165 E. −2.7177

22. Compute $\log_3 19683$.

A. 5 B. 6 C. 7 D. 8 E. 9

23. Use the properties of logarithms to solve the equation for *x*. $\log(x-8) = 2$

A. 98 B. 103 C. 108 D. 113 E. No solution

24. Solve the equation for *x*, to four decimal places, by writing it in logarithmic form, then use the calculator. $9 = e^{x/5}$

A. 10.9832 B. 10.9841 C. 10.9861

D. 10.9872 E. 10.9888

25. A person has 30 coins with a value of $4.30. If the coins are all quarters and nickels, how many of each does the person have?

A. 16 quarters, 14 nickels

B. 16 quarters, 15 nickels

C. 15 quarters, 15 nickels

D. 14 quarters, 16 nickels

E. 12 quarters, 18 nickels

26. Starting with $\begin{bmatrix} 1 & 2 & 1 & 3 \\ 5 & 8 & 7 & 2 \\ 6 & 1 & 3 & 4 \end{bmatrix}$, perform the transformation $-3R_1 + R_2$.

A. $\begin{bmatrix} 1 & -6 & -3 & 3 \\ 2 & 8 & 4 & -7 \\ 6 & 1 & 3 & 4 \end{bmatrix}$

B. $\begin{bmatrix} 1 & 2 & 1 & 3 \\ 2 & 2 & 4 & -7 \\ 6 & 1 & 3 & 4 \end{bmatrix}$

C. $\begin{bmatrix} -3 & -6 & -3 & -9 \\ 5 & 8 & 7 & 2 \\ 6 & 1 & 3 & 4 \end{bmatrix}$

D. $\begin{bmatrix} -3 & -6 & -3 & -9 \\ 2 & 2 & 4 & -7 \\ 6 & 1 & 3 & 4 \end{bmatrix}$

E. $\begin{bmatrix} 2 & 2 & 4 & -7 \\ 6 & 1 & 3 & 4 \end{bmatrix}$

27. Solve (if possible) the system of three equations in three variables, using the method of Gauss-Jordan elimination.

$$2x + 5y + 3z = 18$$
$$-x + y + 3z = 7$$
$$x + 4y - z = 7$$

A. (1, 2, 2)

B. (3, 1, 2)

C. (1, 3, 0)

D. (4, 0, 4)

E. No solution

28. A company makes two different products. Product A requires 20 hours of assembly and 5 pounds of a certain raw material, while product B requires 10 hours of assembly and 10 pounds of the same raw material. There are 1,200 hours of assembly and 1,000 pounds of the raw material available. If the profit from each unit of product A is \$400 and from each unit of product B is \$350, what is the maximum profit the company can make?

A. \$40,000 B. \$21,000 C. \$38,000

D. \$45,000 E. \$25,000

29. Solve the matrix equation $\begin{bmatrix} x & x+z \\ y & y+z \end{bmatrix} = \begin{bmatrix} 7 & 9 \\ 4 & 6 \end{bmatrix}$.

A. (7, 4, 2) B. (7, 4, 9) C. (7, 4, 6)

D. (7, 4, 0) E. (7, 4, 3)

30. Perform the multiplication $\begin{bmatrix} 8 & 4 \\ 9 & 7 \end{bmatrix}\begin{bmatrix} 8 & 7 \\ 3 & 6 \end{bmatrix}$, if possible.

A. $\begin{bmatrix} 76 & 80 \\ 93 & 105 \end{bmatrix}$ B. $\begin{bmatrix} 64 & 28 \\ 27 & 42 \end{bmatrix}$ C. $\begin{bmatrix} 105 & 80 \\ 93 & 76 \end{bmatrix}$

D. $\begin{bmatrix} 80 & 76 \\ 105 & 93 \end{bmatrix}$ E. Not possible

31. Compute the determinant $\begin{vmatrix} 6 & 9 \\ 8 & 9 \end{vmatrix}$.

A. 18 B. −18 C. 17 D. −17 E. 0

32. Evaluate $\sum_{k=1}^{5}(k^2+2)$.

A. 65 B. 66 C. 67 D. 68 E. 69

33. The first term $a_1 = 6$ and the common difference $d = -2$ of an arithmetic sequence are given. Find the first four terms.

A. 6, −4, 2, 0 B. 6, −4, 2, 0

C. 6, 4, 2, 0 D. 6, 8, 10, 12

E. 4, 2, 0, −1

34. The first term of an arithmetic sequence is $a_1 = 8$ and the difference is $d = 5$. Find the 5th term.

A. 13 B. 18 C. 28 D. 33 E. 38

35. The first term and the common ratio of a geometric sequence are $a_1 = 6$ and $r = 7$. Find the 8th term.

A. $6(7)^8$ B. $6(7)^7$ C. $6(7)^9$ D. $(42)^7$ E. $(42)^7$

36. Expand $(2a+b)^5$ using Pascal's triangle.

A. $32a^5 + 16a^4b + 8a^3b^2 + 4a^2b^3 + 2ab^4 + b^5$

B. $a^5 + 5a^4b + 10a^3b^2 + 10a^2b^3 + 5ab^4 + b^5$

C. $32a^5 + 80a^4b + 80a^3b^2 + 40a^2b^3 + 10ab^4 + b^5$

D. $32a^5 - 80a^4b + 80a^3b^2 - 40a^2b^3 + 10ab^4 - b^5$

E. $a^5 - 5a^4b + 10a^3b^2 - 10a^2b^3 + 5ab^4 - b^5$

37. ${}_{20}P_4 =$

A. 116,280 B. 116,200 C. 112,240

D. 112,330 E. 112,480

38. In how many ways can the top three of 80 teams be awarded prizes in an academic competition?

A. 80! B. ${}_{80}P_3$ C. ${}_{80}C_3$

D. ${}_{80}C_0$ E. 20

39. If one card is selected from a deck of 52 cards, what is the probability it is the jack of diamonds?

A. $\frac{1}{13}$ B. $\frac{1}{4}$ C. $\frac{1}{12}$ D. $\frac{1}{52}$ E. 1

40. A box contains 76 good radios and 4 defective radios. What is the probability of getting at most one defective radio if 3 radios are selected?

A. $\frac{{}_{76}C_2 \; {}_4C_1}{{}_{80}C_3} + \frac{{}_{76}C_3}{{}_{80}C_3}$

B. $\frac{{}_{76}C_2 \; {}_4C_1}{{}_{80}C_3}$

C. $\frac{{}_4C_1}{{}_{80}C_3}$

D. $\frac{76}{80}$

E. $\frac{1}{4}$

Test form Final C Date ________ Name ________________________
Course ______ Sec ______ Student Number ________________

Final Exam

1. Find the x- and y-intercepts of the graph of $5x + 10y = 50$ by letting y and then x be zero.

2. Use the point-slope form, $y - y_1 = m(x - x_1)$, with $m = \frac{y_2 - y_1}{x_2 - x_1}$ to find the equation of the line through the points (7, 10) and (21, 60).

3. Find the vertex of $y = 6x^2 + 18x - 3$.

4. Find the points of intersection of $y = 2x^2 - 7$ and $y = -3x^2$.

5. Determine the equation of the circle having center (8, −12) and radius 6.

6. Solve the quadratic equation. $x^2 + 20x + 64 = 0$

7. Use the equation $s = -16t^2 + vt$ that gives the height s feet at time t seconds of an object fired upward with a velocity v feet per second. If the object is propelled at 288 feet per second, when will it come back to the ground?

8. Solve the equation $\sqrt{x-3} = 7$.

9. Solve the linear inequality. Write the solution in interval notation.
   $7x + 4 > x + 46$

10. Solve the inequality. Write the solution in interval notation. $|x| < 32$

11. Determine the domain and range of $f(x) = x^2 + 7$.

12. Determine the domain of $f(x) = \frac{(20x + 12)(x - 4)}{x - 4}$.

13. $f(x) = x^4$ and $g(x) = x^2 - 2x$. Find $(g \circ f)(x)$.

14. Of $f(x), g(x)$ and $h(x)$, list all one-to-one functions.

$f(x)$ 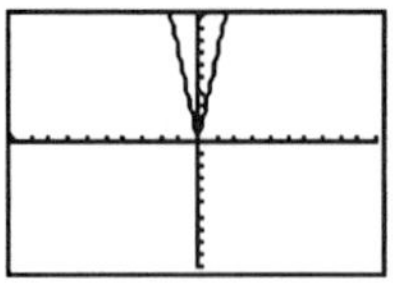

$g(x)$ 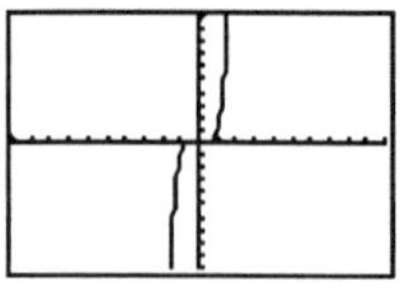

$h(x)$ 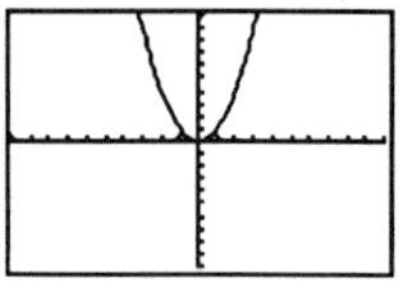

15. Find the inverse function for $f(x) = 3x^7 - 4$.

16. Compute the product $(7+2i)(6+3i)$.

17. Perform the division $\dfrac{7+3i}{2+5i}$.

18. Perform the additions and subtractions. Write the answer in the standard form $a+bi$. $(7+4i)-(9+3i)+(11-7i)$

19. Find the real zeros of $f(x) = x^4 - 8x^3 + 11x^2 + 32x - 60$.

20. Solve using algebraic methods. $9^{x-8} = 9^{5x}$

21. Solve the equation (if possible) to four decimal places. $3^{3x+7} = 6$

22. Compute $\log_4 16384$.

23. Use the properties of logarithms to solve the equation for x. $\log(x-5) = 1$

24. Solve the equation for x, to four decimal places, by writing it in logarithmic form, then use the calculator. $10 = e^{x/7}$

25. A person has 30 coins with a value of \$3.50. If the coins are all quarters and nickels, how many of each does the person have?

26. Starting with $\begin{bmatrix} 1 & 2 & 2 & 1 \\ 7 & 6 & 8 & 9 \\ 1 & 1 & 3 & 1 \end{bmatrix}$, perform the transformation $-3R_1 + R_2$.

27. Solve (if possible) the system of three equations in three variables, using the method of Gauss-Jordan elimination.

$$x+2y+z=7$$
$$5x+y+z=9$$
$$2x-y+3z=6$$

28. A company makes two different products. Product A requires 20 hours of assembly and 5 pounds of a certain raw material, while product B requires 10 hours of assembly and 10 pounds of the same raw material. There are 1,200 hours of assembly and 1,000 pounds of the raw material available. If the profit from each unit of product A is $700 and from each unit of product B is $800, what is the maximum profit the company can make?

29. Solve the matrix equation $\begin{bmatrix} x & x+z \\ y & y+z \end{bmatrix} = \begin{bmatrix} 5 & 7 \\ 7 & 9 \end{bmatrix}$.

30. Perform the multiplication $\begin{bmatrix} 8 & 8 \\ 9 & 7 \end{bmatrix}\begin{bmatrix} 9 & 2 \\ 9 & 1 \end{bmatrix}$, if possible.

31. Compute the determinant $\begin{vmatrix} 2 & 3 \\ 4 & 5 \end{vmatrix}$

32. Evaluate $\sum_{k=1}^{5}(k^2+5)$.

33. The first term $a_1 = 8$ and the common difference $d = -3$ of an arithmetic sequence are given. Find the first four terms.

34. The first term of an arithmetic sequence is $a_1 = 6$ and the difference is $d = 5$. Find the 7th term.

35. The first term and the common ratio of a geometric sequence are $a_1 = 9$ and $r = 5$. Find the 8th term.

36. Expand $(a-2b)^5$ using Pascal's triangle.

37. $_{30}P_6 =$

38. In how many ways can the top three of 90 teams be awarded prizes in an academic competition?

39. If one card is selected from a deck of 52 cards, what is the probability it is the ten of clubs?

40. A box contains 96 good radios and 4 defective radios. What is the probability of getting at most one defective radio if 3 radios are selected?

Test form Final D Date ________ Name ______________________

Course ______ Sec ______ Student Number ______________

Final Exam

1. Find the x- and y-intercepts of the graph of $6x+9y=36$ by letting y and then x be zero.

2. Use the point-slope form, $y-y_1=m(x-x_1)$, with $m=\frac{y_2-y_1}{x_2-x_1}$ to find the equation of the line through the points (3, 10) and (9, 40).

3. Find the vertex of $y=8x^2+24x-6$.

4. Find the points of intersection of $y=4x^2-8$ and $y=-3x^2$.

5. Determine the equation of the circle having center (9, –11) and radius 7.

6. Solve the quadratic equation. $x^2-20x+64=0$

7. Use the equation $s=-16t^2+vt$ that gives the height s feet at time t seconds of an object fired upward with a velocity v feet per second. If the object is propelled at 320 feet per second, when will it come back to the ground?

8. Solve the equation $\sqrt{x-7}=9$.

9. Solve the linear inequality. Write the solution in interval notation.
$9x+5>2x+68$

10. Solve the inequality. Write the solution in interval notation. $|x|<51$

11. Determine the domain and range of $f(x)=x^2+4$.

12. Determine the domain of $f(x)=\dfrac{(36x+7)(x-5)}{x-5}$.

13. $f(x)=x^5$ and $g(x)=x^3+3x$. Find $(g\circ f)(x)$.

14. Of $f(x), g(x)$ and $h(x)$, list all one-to-one functions.

$f(x)$ $g(x)$ $h(x)$

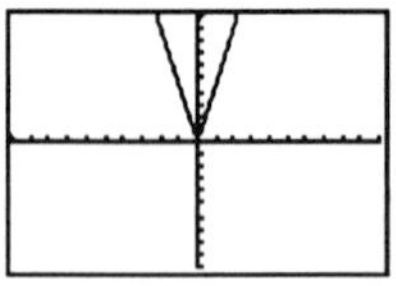
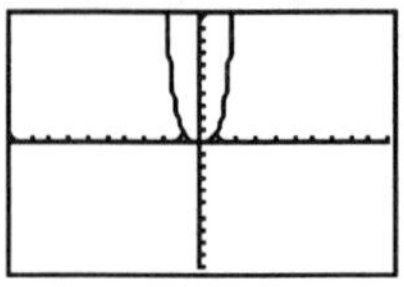
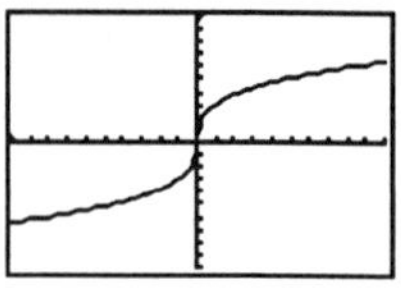

15. Find the inverse function for $f(x) = 5x^{11} - 6$.

16. Compute the product $(5+3i)(2+4i)$.

17. Perform the division $\dfrac{3+7i}{5+2i}$.

18. Perform the additions and subtractions. Write the answer in the standard form $a+bi$. $(5+7i)-(6+8i)+(12-3i)$

19. Find the real zeros of $f(x) = x^4 - 9x^3 + 14x^2 + 36x - 72$.

20. Solve using algebraic methods. $7^{x-6} = 7^{4x}$

21. Solve the equation (if possible) to four decimal places. $6^{4x+3} = 8$

22. Compute $\log_4 4096$.

23. Use the properties of logarithms to solve the equation for x. $\log(x+3) = 1$

24. Solve the equation for x, to four decimal places, by writing it in logarithmic form, then use the calculator. $9 = e^{x/6}$

25. A person has 32 coins with a value of \$3.60. If the coins are all quarters and nickels, how many of each does the person have?

26. Starting with $\begin{bmatrix} 1 & 2 & 2 & 1 \\ 8 & 7 & 5 & 6 \\ 1 & 4 & 1 & 3 \end{bmatrix}$, perform the transformation $-3R_1 + R_2$.

27. Solve (if possible) the system of three equations in three variables, using the method of Gauss-Jordan elimination.

$$4x + y + 3z = 13$$
$$3x + 3y + 2z = 14$$
$$x + 2y + 4z = 10$$

28. A company makes two different products. Product A requires 20 hours of assembly and 5 pounds of a certain raw material, while product B requires 10 hours of assembly and 10 pounds of the same raw material. There are 1,200 hours of assembly and 1,000 pounds of the raw material available. If the profit from each unit of product A is $2,100 and from each unit of product B is $2,400, what is the maximum profit the company can make?

29. Solve the matrix equation $\begin{bmatrix} x & x+z \\ y & y+z \end{bmatrix} = \begin{bmatrix} 3 & 7 \\ 2 & 6 \end{bmatrix}$.

30. Perform the multiplication $\begin{bmatrix} 7 & 9 \\ 9 & 6 \end{bmatrix}\begin{bmatrix} 3 & 4 \\ 5 & 6 \end{bmatrix}$, if possible.

31. Compute the determinant $\begin{vmatrix} 9 & 5 \\ 9 & 4 \end{vmatrix}$.

32. Evaluate $\sum_{k=1}^{5} (k^2 + 9)$.

33. The first term $a_1 = 7$ and the common difference $d = -5$ of an arithmetic sequence are given. Find the first four terms.

34. The first term of an arithmetic sequence is $a_1 = 8$ and the difference is $d = 4$. Find the 7th term.

35. The first term and the common ratio of a geometric sequence are $a_1 = 8$ and $r = 9$. Find the 8th term.

36. Expand $(2a - b)^5$ using Pascal's triangle.

37. ${}_{30}P_4 =$

38. In how many ways can the top three of 78 teams be awarded prizes in an academic competition?

39. If one card is selected from a deck of 52 cards, what is the probability it is the eight of hearts?

40. A box contains 95 good radios and 5 defective radios. What is the probability of getting at most one defective radio if 3 radios are selected?

SOLUTIONS

CHAPTER R[1]

FORM A

1. [R.1/4/3] C

2. [R.1/10/3] B

3. [R.1/11/2] B

4. [R.2/20/1] B

5. [R.2/22/1] A

6. [R.2/23/1] B

7. [R.2/21/3] C

8. [R.3/31/3] C

9. [R.3/32/2] C

10.[R.3/32/2] A

11.[R.3/35/1] B

12.[R.4/38/1] C

13.[R.4/40/2] B

14.[R.4/40/2] A

15.[R.545/2] A

16.[R.5/46/2] E

17.[R.5/46/2] A

18.[R.5/47/1] C

19.[R.5/52/3] B

20.[R.6/52/2] A

21.[R.6/54/2] E

22.[R.6/67/3] B

23.[R.6/67/1] A

24.[R.6/69/2] B

25.[R.6/73/1] E

FORM B

1. [R.1/4/3] A

2. [R.1/10/3] B

3. [R.1/11/2] D

4. [R.2/20/1] C

5. [R.2/22/1] A

6. [R.2/23/1] B

7. [R.2/21/3] C

8. [R.3/31/3] B

9. [R.3/32/2] C

10.[R.3/32/2] E

11.[R.3/35/1] A

12.[R.4/38/1] A

13.[R.4/40/2] A

14.[R.4/40/2] B

15.[R.5/45/2] B

16.[R.5/46/2] D

17.[R.5/46/2] A

18.[R.5/47/1] C

19.[R.5/52/3] A

20.[R.6/52/2] A

21.[R.6/54/2] E

22.[R.6/67/3] C

23.[R.6/67/1] A

24.[R.6/69/2] E

25.[R.6/73/1] E

FORM C

1. [R.1/4/3] B

2. [R.1/10/3] B

3. [R.1/11/2] C

4. [R.2/20/1] A

5. [R.2/22/1] A

6. [R.2/23/1] A

7. [R.2/21/3] A

8. [R.3/31/3] D

9. [R.3/32/2] C

10.[R.3/32/2] D

11.[R.3/35/1] B

12.[R.4/38/1] B

[1] [x/y/z] is [sec./pg./diff.] where diff. 1 is hardest.

13.[R.4/40/2] B

14.[R.4/40/2] D

15.[R.5/45/2] B

16.[R.5/46/2] A

17.[R.5/46/2] A

18.[R.5/47/1] C

19.[R.5/52/3] B

20.[R.6/52/2] A

21.[R.6/54/2] E

22.[R.6/67/3] A

23.[R.6/67/1] A

24.[R.6/69/2] C

25.[R.6/73/1] E

FORM D

1. [R.1/4/3]

$-11, -\frac{3}{7}, 0$

2. [R.1/10/3]

Commutative property

of addition

3. [R.1/11/2] 0.1393

4. [R.2/20/1] 6

5. [R.2/22/1] $\frac{27y^2}{x^4}$

6. [R.2/23/1]

$|xy| \sqrt[6]{3y^4}$

7. [R.2/21/3] 0.0392

8. [R.3/31/3] 8

9. [R.3/32/2]

$7x^2 + 5x + 5$

10.[R.3/32/2]

$35x^4 + 15x^3 + 28x + 12$

11.[R.3/35/1]

$x + 8 + \frac{-7}{x+2}$

12.[R.4/38/1]

$(x-5)(x+8)$

13.[R.4/40/2]

$(9x - 5y - 3)(9x + 5y + 3)$

14.[R.4/40/2]

$(2x-7)^2$

15.[R.5/45/2] $\frac{x+7}{x+4}$

16.[R.5/46/2]

$\frac{2(5x+9)}{9}$

17.[R.5/46/2] 1

18.[R.5/47/1]

$\frac{31x-3}{(x-1)(3x+1)}$

19.[R.5/52/3] 2

20.[R.6/52/2] $\frac{1}{3}$

21.[R.6/54/2]

No solution

22.[R.6/67/3] 59

23.[R.6/67/1] $(5, \infty)$

24.[R.6/69/2] 22.7349

25.[R.6/73/1]

$10x + 5h$

FORM E

1. [R.1/4/3]

$-15, -\frac{2}{5}, 0$

2. [R.1/10/3]

Commutative property

of addition

3. [R.1/11/2] 0.5710

4. [R.2/20/1] 3

5. [R.2/22/1] $\frac{27y^4}{x^2}$

6. [R.2/23/1]

$|xy|\sqrt[4]{5y^2}$

7. [R.2/21/3] 0.0014

8. [R.3/31/3] 8

9. [R.3/32/2]

$3x^2+3x+4$

10.[R.3/32/2]

$8x^4+20x^3+36x+90$

11.[R.3/35/1]

$x+10+\frac{-19}{x+2}$

12.[R.4/38/1]

$(x+7)(x-6)$

13.[R.4/40/2]

$(8x-7y+4)(8x+7y-4)$

14.[R.4/40/2]

$(3x-7)^2$

15.[R.5/45/2] $\frac{x+8}{x+4}$

16.[R.5/46/2]

$\frac{2(4x+11)}{3}$

17.[R.5/46/2] 1

18.[R.5/47/1]

$\frac{25x+3}{(x-1)(3x+1)}$

19.[R.5/52/3] 2

20.[R.6/52/2] $\frac{1}{3}$

21.[R.6/54/2]

No solution

22.[R.6/67/3] 41

23.[R.6/67/1] $(13, \infty)$

24.[R.6/69/2] 20.5009

25.[R.6/73/1]

$12x+6h$

FORM F

1. [R.1/4/3]

$-13, -\frac{4}{7}, 0$

2. [R.1/10/3]

Commutative property

of addition

3. [R.1/11/2] 1.7790

4. [R.2/20/1] 9

5. [R.2/22/1] $\frac{27y^3}{x^4}$

6. [R.2/23/1]

$|xy|\sqrt[8]{11y^2}$

7. [R.2/21/3] 0.0128

8. [R.3/31/3] 7

9. [R.3/32/2]

$4x^2+3x+4$

10.[R.3/32/2]

$30x^4+40x^3+42x+56$

11.[R.3/35/1]

$x+6+\frac{-3}{x+2}$

12.[R.4/38/1]

$(x+8)(x-2)$

13.[R.4/40/2]

$(4x-9y-1)(4x+9y+1)$

14.[R.4/40/2]

$(4x-5)^2$

15.[R.5/45/2] $\frac{x+9}{x+5}$

16.[R.5/46/2]

$\frac{2(6x+11)}{11}$

17.[R.5/46/2] 1

18.[R.5/47/1]

$\dfrac{11x+1}{(x-1)(3x+1)}$

19.[R.5/52/3] 2

20.[R.6/52/2] $\dfrac{1}{3}$

21.[R.6/54/2]

No solution

22.[R.6/67/3] 17

23.[R.6/67/1] $(6, \infty)$

24.[R.6/69/2] 17.3304

25.[R.6/73/1]

$18x+9h$

CHAPTER 1

FORM A

1. [1.1/89/3] D

2. [1.1/92/3] C

3. [1.2/97/2] B

4. [1.2/98/2] A

5. [1.2/100/3] B

6. [1.2/100/3] A

7. [1.2/104/1] A

8. [1.2/102/1] D

9. [1.2/106/1] E

10.[1.3/114/3] A

11.[1.3/114/2] A

12.[1.3/115/1] C

13.[1.4/124/1] B

14.[1.4/124/2] C

15.[1.4/125/2] B

16.[1.4/126/2] D

17.[1.4/126/2] C

18.[1.4/127/1] A

19.[1.5/130/1] B

20.[1.5/134/2] B

21.[1.6/141/3] B

22.[1.6/142/3] D

23.[1.6/144/2] B

24.[1.6/145/2] B

25.[1.6/147/1] D

FORM B

1. [1.1/89/3] E

2. [1.1/92/3] C

3. [1.2/97/2] D

4. [1.2/98/2] B

5. [1.2/100/3] A

6. [1.2/100/3] B

7. [1.2/104/1] B

8. [1.2/102/1] E

9. [1.2/106/1] C

10.[1.3/114/3] D

11.[1.3/114/2] A

12.[1.3/115/1] C

13.[1.4/124/1] D

14.[1.4/124/2] C

15.[1.4/125/3] C

16.[1.4/126/2] A

17.[1.4/126/2] E

18.[1.4/127/1] E

19.[1.5/130/1] A

20.[1.5/134/2] B

21.[1.6/141/3] B

22.[1.6/142/3] E

23.[1.6/144/2] C

24.[1.6/145/2] D

25.[1.6/147/1] D

FORM C

1. [1.1/89/3] B

2. [1.1/92/3] E

3. [1.2/97/2] B

4. [1.2/98/2] A

5. [1.2/100/3] C

6. [1.2/100/3] A

7. [1.2/104/1] D

8. [1.2/102/1] B

9. [1.2/106/1] E

10.[1.3/114/3] D

11.[1.3/114/2] A

12.[1.3/115/1] C

13.[1.4/124/1] A

14.[1.4/124/2] D

15.[1.4/125/3] B

16.[1.4/126/2] C

17.[1.4/126/2] A

18.[1.4/127/1] C

19.[1.5/130/1] E

20.[1.5/134/2] B

21.[1.6/141/3] D

22.[1.6/142/3] C

23.[1.6/144/2] A

24.[1.6/145/2] C

25.[1.6/147/1] D

FORM D

1. [1.1/89/3]

(0, 6) and (1, 15)

2. [1.2/92/3]

x-intercept (2, 0),
y-intercept (0, –12)

3. [1.2/97/2] 2

4. [1.2/98/2]

$y-9=-\frac{7}{2}(x-4)$ or

$y-2=-\frac{7}{2}(x-6)$

5. [1.2/100/3] $y=1$

6. [1.2/100/3] $x=2$

7. [1.2/1046/1]

$y-1=4(x-2)$

8. [1.2/102/1]

$y-7=-\frac{1}{4}(x-5)$

9. [1.2/106/1]

$y=3.1x-1$

10.[1.3/114/3] –25.78

11.[1.3/114/2] 25

12.[1.3/115/1]

$y=67x+935$

13.[1.4/124/1] (–1, –10)

14.[1.4/124/2] $x=-1$

15.[1.4/125/3] (0, –7)

16.[1.4/126/2]

(3, 0) and (5, 0)

17.[1.4/126/2]

(–0.8697, 0) and
(2.0697, 0)

18.[1.4/127/1] 5

19.[1.5/130/1] 320 m/s

20.[1.5/134/2]

(–0.5201, 5.6283) and
(1.6966, 4.4548)

21.[1.6/141/3] 6.0828

22.[1.6/142/3] (–2, 4.5)

23.[1.6/144/2]

Center: (–6, 7),
Radius: 5

24.[1.6/145/2]

$(x+7)^2+(y-5)^2=16$

25.[1.6/147/1]

Center: (–1, –1.5),
Radius: 4.27

FORM E

1. [1.1/89/3]

(0, 1) and (1, –1)

2. [1.1/92/3]

x-intercept (–6, 0),
y-intercept (0, –18)

3. [1.2/97/2] 2

4. [1.2/98/2]

$y-8=\frac{5}{3}(x-7)$ or

$y-3=\frac{5}{3}(x-4)$

5. [1.2/100/3] $y=4$

6. [1.2/100/3] $x=6$

7. [1.2/104/1]

$y-9=5(x-3)$

8. [1.2/102/1]

$y-6=-\frac{1}{5}(x-2)$

9. [1.2/106/1]

$y=5.5x+2.5$

10.[1.3/114/3] –59.28

11.[1.3/114/2] 30

12.[1.3/115/1]

$y=60x+981$

13.[1.4/124/1]

(–1.125, –6.0625)

14.[1.4/124/2]

$x=-1.125$

15.[1.4/125/3] (0, –1)

16.[1.4/126/2]

(3, 0) and (7, 0)

17.[1.4/126/2]

(–1.7787, 0) and
(4.7787, 0)

18.[1.4/127/1] 7

19.[1.5/130/1] 384 m/s

20.[1.5/134/2]

(–0.6988, 6.7395) and
(2.3205, 4.1283)

21.[1.6/141/3] 16.0312

22.[1.6/142/3] (–1, 10.5)

23.[1.6/144/2]

Center: (–9, 3),
Radius: 7

24.[1.6/145/2]

$(x+8)^2+(y-2)^2=81$

25.[1.6/147/1]

Center: (–3, –2),
Radius: 7.42

FORM F

1. [1.1/89/3]

(0, 8) and (1, 13)

2. [1.1/92/3]

x-intercept (4, 0),
y-intercept (0, 12)

3. [1.2/97/2] 5

4. [1.2/98/2]

$y-8=-\frac{3}{2}(x-1)$ or

$y-2=-\frac{3}{2}(x-5)$

5. [1.2/100/3] $y=7$

6. [1.2/100/3] $x=4$

7. [1.2/104/1]

$y-9=-7(x-4)$

8. [1.2/102/1]

$y-6=\frac{1}{7}(x-2)$

9. [1.2/106/1]

$y = 2.1 + 3$

10.[1.3/114/3] –23.81

11.[1.3/114/2] 20

12.[1.3/115/1]

$y = 61x + 972$

13.[1.4/124/1]

(–1.25, –9.125)

14.[1.4/124/2] $x = -1.25$

15.[1.4/125/3] (0, –6)

16.[1.4/126/2]

(2, 0) and (8, 0)

17.[1.4/126/2]

(–1.6085, 0) and
(3.1085, 0)

18.[1.4/127/1] 3

19.[1.5/130/1] 288 m/s

20.[1.5/134/2]

(–0.753, 6.037) and
(1.886, 4.902)

21.[1.6/141/3] 8.9443

22.[1.6/142/3] (–1.5, 5)

23.[1.6/144/2]

Center: (–3, 8),
Radius: 6

24.[1.6/145/2]

$(x+9)^2 + (y-2)^2 = 64$

25.[1.6/147/1]

Center: $\left(-\frac{3}{2}, -2\right)$,
Radius: 6.0208

CHAPTER 2

FORM A

1. [2.1/165/3] C

2. [2.1/165/2] A

3. [2.1/166/3] A

4. [2.1/170/2] A

5. [2.1/172/3] C

6. [2.2/175/2] D

7. [2.2/176/1] E

8. [2.3/183/1] C

9. [2.3/187/2] E

10.[2.3/184/1] C

11.[2.3/184/1] E

12.[2.3/185/2] B

13.[2.3/188/3] A

14.[2.4/192/3] A

15.[2.4/192/3] B

16.[2.4/192/2] E

17.[2.4/193/2] D

18.[2.4/195/1] B

19.[2.4/195/1] A

20.[2.5/205/3] A

21.[2.5/205/2] D

22.[2.5/207/2] B

23.[2.5/207/2] A

24.[2.5/207/2] B

25.[2.5/207/2] C

FORM B

1. [2.1/165/3] D

2. [2.1/165/2] B

3. [2.1/166/3] B

4. [2.1/170/2] D

5. [2.1/172/3] C

6. [2.2/175/2] C

7. [2.2/176/1] D

8. [2.3/183/1] E

9. [2.3/187/2] B

10.[2.3/184/1] A

11.[2.3/184/1] E

12.[2.3/185/2] B

13.[2.3/188/3] D

14.[2.4/192/3] A

15.[2.4/192/3] D

16.[2.4/192/2] E

17.[2.4/193/2] A

18.[2.4/195/1] B

19.[2.4/195/1] A

20.[2.5/205/3] A

21.[2.5/205/2] C

22.[2.5/207/2] B

23.[2.5/207/2] D

24.[2.5/207/2] A

25.[2.5/207/2] D

FORM C

1. [2.1/165/3] A

2. [2.1/165/2] C

3. [2.1/166/3] B

4. [2.1/170/2] B

5. [2.1/172/3] C

6. [2.2/175/2] A

7. [2.2/176/1] C

8. [2.3/183/1] B

9. [2.3/187/2] A

10.[2.3/184/1] B

11.[2.3/184/1] E

12.[2.3/185/2] C

13.[2.3/188/3] B

14.[2.4/192/3] C

15.[2.4/192/3] E

16.[2.4/192/2] C

17.[2.4/193/2] A

18.[2.4/195/1] B

19.[2.4/195/1] A

20.[2.5/205/3] A

21.[2.5/205/2] B

22.[2.5/207/2] A

23.[2.5/207/2] D

24.[2.5/207/2] D

25.[2.5/207/2] E

FORM D

1. [2.1/165/3]

$x = -21, -2$

2. [2.1/165/2]

$x = -\frac{1}{5}, -8$

3. [2.1/166/3]

$-2.4495, 2.4495$

4. [2.1/170/2]

$-1.3423, 4.8423$

5. [2.1/172/3]

No real solutions

6. [2.2/175/2]

6 seconds

7. [2.2/176/1]

13 feet by 16 feet

8. [2.3/183/1]

$-1, 1, -2, 2$

9. [2.3/187/2] 625

10.[2.3/184/1]

$-\frac{1}{3}, -\frac{1}{10}$

11.[2.3/184/1]

No solution

12.[2.3/185/2] 12

13.[2.3/188/3] $-2, 0, 4$

14.[2.4/192/3] $\left(-\infty, \frac{2}{5}\right)$

15.[2.4/192/3] $(2, \infty)$

16.[2.4/192/2] $(-2, \infty)$

17.[2.4/193/2] $[1, 8]$

18.[2.4/195/1] $(-7, -3)$

19.[2.4/195/1] $[-19, -4]$

20.[2.5/205/3] $-27, 27$

21.[2.5/205/2] $-24, 16$

22.[2.5/207/2] $(-13, 13)$

23.[2.5/207/2]

$(-\infty, -23] \cup [23, \infty)$

24.[2.5/207/2] $(-8, 4)$

25.[2.5/207/2]

$(-\infty, -8) \cup (4, \infty)$

FORM E

1. [2.1/165/3]

$x = -10, -3$

2. [2.1/165/2]

$x = -\frac{1}{5}, -9$

3. [2.1/166/3]

$-3.7417, 3.7417$

4. [2.1/170/2]

$-1.2404, 5.2404$

5. [2.1/172/3]

One real solution

6. [2.2/175/2]

7 seconds

7. [2.2/176/1]

23 feet by 20 feet

8. [2.3/183/1]

$-1, 1, -3, 3$

9. [2.3/187/2] 16

10.[2.3/184/1]

$-\frac{1}{5}, -\frac{1}{10}$

11.[2.3/184/1]

No solution

12.[2.3/185/2] 17

13.[2.3/188/3] 0, 1, 7

14.[2.4/192/3] $(-\infty, -1)$

15.[2.4/192/3] $(2, \infty)$

16.[2.4/192/2] $(-2, \infty)$

17.[2.4/193/2] $[-1, 4]$

18.[2.4/195/1] $(-11, -2)$

19.[2.4/195/1]

$(-\infty, -20] \cup [5, \infty)$

20.[2.5/205/3] $-29, 29$

21.[2.5/205/2] $-32, 28$

22.[2.5/207/2] $(-60, 60)$

23.[2.5/207/2]

$(-\infty, -10] \cup [10, \infty)$

24.[2.5/207/2] $(-52, 48)$

25.[2.5/207/2]

$(-\infty, -5) \cup (1, \infty)$

FORM F

1. [2.1/165/3] $-10, -5$

2. [2.1/165/2]

$x = -\frac{1}{3}, -8$

3. [2.1/1661/3]

$-1.7321, 1.7321$

4. [2.1/170/2]

$-1.3646, 3.1146$

5. [2.1/172/3]

Two real solutions

6. [2.2/175/2]

8 seconds

7. [2.2/176/1]

22 feet by 25 feet

8. [2.3/183/1]

$-1, 1, -4, 4$

9. [2.3/187/2] 81

10.[2.3/184/1]

$-\frac{1}{8}, -\frac{1}{10}$

11.[2.3/184/1]

No solution

12.[2.3/185/2] 5

13.[2.3/188/3] –7, 0, 5

14.[2.4/192/3] $(-\infty, 1)$

15.[2.4/192/3] $(2, \infty)$

16.[2.4/192/2] $(-2, \infty)$

17.[2.4/193/2] $\left[\frac{1}{2}, \frac{11}{2}\right]$

18.[2.4/195/1] $(-16, -3)$

19.[2.4/195/1]

$(-\infty, -15] \cup [-4, \infty)$

20.[2.5/205/3] –32, 32

21.[2.5/205/2] –10, 8

22.[2.5/207/2] $(-47, 47)$

23.[2.5/207/2]

$(-\infty, -52] \cup [52, \infty)$

24.[2.5/207/2] $(-21, 17)$

25.[2.5/207/2]

$(-\infty, -23) \cup (19, \infty)$

CHAPTER 3

FORM A

1. [3.1/221/3] C

2. [3.1/224/3] C

3. [3.2/228/2] A

4. [3.2/228/3] E

5. [3.2/229/3] C

6. [3.3/240/3] B

7. [3.3/245/2] C

8. [3.3/245/2] A

9. [3.4/259/1] B

10.[3.4/259/1] E

11.[3.4/258/3] D

12.[3.5/275/1] C

13.[3.5/281/2] A

14.[3.5/281/2] D

15.[3.5/282/2] A

16.[3.5/283/1] D

17.[3.6/287/3] B

18.[3.6/289/1] C

19.[3.6/291/1] E

20.[3.6/290/1] A

21.[3.7/306/2] D

22.[3.7/307/2] D

23.[3.7/308/1] D

24.[3.7/308/1] D

25.[3.7/309/1] B

FORM B

1. [3.1/221/3] A

2. [3.1/224/3] E

3. [3.2/228/2] C

4. [3.2/228/3] B

5. [3.2/229/3] C

6. [3.3/240/3] A

7. [3.3/245/2] C

8. [3.3/245/2] B

9. [3.4/259/1] B

10.[3.4/259/1] D

11.[3.4/258/3] A

12.[3.5/275/1] A

13.[3.5/281/2] A

14.[3.5/281/2] A

15.[3.5/282/2] C

16.[3.5/283/1] D

17.[3.6/287/3] C

18.[3.6/289/1] B

19.[3.6/291/1] D

20.[3.6/290/1] D

21.[3.7/306/2] D

22.[3.7/307/2] D

23.[3.7/308/1] C

24.[3.7/308/1] E

25.[3.7/309/1] E

FORM C

1. [3.1/221/3] D
2. [3.1/224/3] D
3. [3.2/228/2] C
4. [3.2/228/3] C
5. [3.2/229/3] A
6. [3.3/240/3] D
7. [3.3/245/2] C
8. [3.3/245/2] E
9. [3.4/259/1] D

10.[3.4/259/1] A

11.[3.4/258/3] B

12.[3.5/275/1] B

13.[3.5/281/2] C

14.[3.5/281/2] C

15.[3.5/282/2] C

16.[3.5/283/1] D

17.[3.6/287/3] A

18.[3.6/289/1] B

19.[3.6/291/1] E

20.[3.6/290/1] C

21.[3.7/306/2] D

22.[3.7/307/2] D

23.[3.7/308/1] B

24.[3.7/308/1] E

25.[3.7/309/1] A

FORM D

1. [3.1/221/3] –9, 0, 2

2. [3.1/224/3]

$f(x) = 2x^3 - 30x^2 + 52x + 480$

3. [3.2/228/2]

$\left(-\infty, \frac{13}{2}\right)$

4. [3.2/228/3] $-\frac{25}{4}$

5. [3.2/229/3] 18.1292

6. [3.3/240/3]

Domain: $(-\infty, \infty)$;
Range: $[0, \infty)$

7. [3.3/245/2]

$g(x)$ and $h(x)$ only

8. [3.3/245/2]

Symmetric about the origin

9. [3.4/259/1]

Horizontal shift of 9 units to the left, reflection about the x-axis, stretching by a factor of 7, and vertical shift of 8 units down

10.[3.4/259/1]

$(x-1)^2 - 10$

11.[3.4/258/3]

$f(x) = |x-3| - 7$

12.[3.5/275/1]

All real numbers except 2

13.[3.5/281/2]

$x = -\frac{9}{4}$

14.[3.5/281/2]

$y = \frac{1}{7}$

15.[3.5/282/2]

$x = -\frac{7}{5}$

16.[3.5/283/1]

Slant asymptote $y = x - 5$; vertical asymptote $x = -2$

17.[3.6/287/3]

$\sqrt{x-7} + x + 2$

18.[3.6/289/1]

$\frac{-3}{x^2 + 28}$

19.[3.6/291/1]

$-\frac{x^3}{100} + 3x^2 + 5000x - 5000$

20.[3.6/290/1] 344

21.[3.7/306/2]

$g(x)$ and $h(x)$

22.[3.7/307/2]

$g(x)$ and $h(x)$

23.[3.7/308/1]

$f^{-1}(x) = \frac{x+25}{6}$

24.[3.7/308/1]

$f^{-1}(x) = \sqrt[7]{x-8}$

25.[3.7/309/1] $\sqrt[5]{7v}$

FORM E

1. [3.1/221/3]

2. [3.1/224/3]

$f(x) = 4x^3 - 60x^2 + 104x + 960$

3. [3.2/228/2] −12, 0, 5

$\left(-\infty, \frac{17}{2}\right)$

4. [3.2/228/3] 13.75

5. [3.2/229/3] 10.1986

6. [3.3/240/3]

Domain: $(-\infty, \infty)$;
Range: $[0, \infty)$

7. [3.3/245/2]

$g(x)$ and $h(x)$ only

8. [3.3/245/2]

Symmetric about the origin

9. [3.4/259/1]

Horizontal shift of 10 units to the left, reflection about the x-axis, stretching by a factor of 3, and vertical shift of 8 units down

10.[3.4/259/1]

$(x-1)^2 - 20$

11.[3.4/258/3]

$f(x) = -|x+3| + 2$

12.[3.5/275/1]

All real numbers except 7

13.[3.5/281/2] $x = \frac{3}{2}$

14.[3.5/281/2] $y = \frac{6}{7}$

15.[3.5/282/2] $x = \frac{7}{5}$

16.[3.5/283/1]

Slant asymptote $y = x + 6$; vertical asymptote $x = -3$

17.[3.6/287/3]

$2\sqrt{x-7} + x + 4$

18.[3.6/289/1]

$\frac{-10}{x^2+2}$

19.[3.6/291/1]

$-\frac{x^3}{100}+3x^2+5000x-5000$

20.[3.6/290/1] 730

21.[3.7/306/2]

$f(x)$ and $h(x)$

22.[3.7/307/2]

$f(x)$ and $h(x)$

23.[3.7/308/1]

$f^{-1}(x)=\frac{x-7}{10}$

24.[3.7/308/1]

$f^{-1}(x)=\sqrt[3]{x-26}$

25.[3.7/309/1] $\sqrt[11]{4v}$

FORM F

1. [3.1/221/3] –8, 0, 7

2. [3.1/224/3]

$f(x)=3x^3-45x^2+78x+720$

3. [3.2/228/2]

$\left(-\infty, \frac{15}{2}\right)$

4. [3.2/228/3] 46

5. [3.2/229/3] 12.0347

6. [3.3/240/3]

Domain: $(-\infty, \infty)$;
Range: $[0, \infty)$

7. [3.3/245/2]

$g(x)$ and $h(x)$ only

8. [3.3/245/2]

Symmetric about the y-axis

9. [3.4/259/1]

Horizontal shift of 10 units to the left, reflection about the x-axis, stretching by a factor of 8, and vertical shift of 7 units down

10.[3.4/259/1]

$(x-6)^2-45$

11.[3.4/258/3]

$f(x)=|x+4|-7$

12.[3.5/275/1]

All real numbers except 5

13.[3.5/281/2] $x=-\frac{15}{4}$

14.[3.5/281/2] $y=\frac{3}{7}$

15.[3.5/282/2] $x=-\frac{7}{9}$

16.[3.5/283/1]

Slant asymptote $y=x-6$; vertical asymptote $x=-12$

17.[3.6/287/3]

$\sqrt{x-12}+x+5$

18.[3.6/289/1]

$\frac{3}{x^2+16}$

19.[3.6/291/1]

$-\frac{x^3}{100}+3x^2+5000x-5000$

20.[3.6/290/1] 513

21.[3.7/306/2]

$g(x)$ and $h(x)$

22.[3.7/307/2]

$g(x)$ and $h(x)$

23.[3.7/308/1]

$f^{-1}(x)=\frac{x+21}{3}$

24.[3.7/308/1]

$f^{-1}(x) = \sqrt[9]{x-23}$

25.[3.7/309/1]

$\sqrt[9]{8v}$

CHAPTER 4

FORM A

1. [4.1/327/3] A
2. [4.1/327/2] B
3. [4.1/327/3] C
4. [4.1/329/3] A
5. [4.2/331/3] C
6. [4.2/332/2] E
7. [4.2/335/2] C
8. [4.2/335/2] A
9. [4.2/333/1] A
10. [4.2/335/1] E
11. [4.2/336/2] E
12. [4.2/334/3] A
13. [4.3/341/1] A
14. [4.3/342/1] E
15. [4.3/342/1] B
16. [4.3/344/3] B
17. [4.3/344/2] A
18. [4.3/343/2] E
19. [4.3/344/2] C
20. [4.3/345/1] A
21. [4.3/345/1] C
22. [4.4/348/2] A
23. [4.4/350/1] C
24. [4.4/352/1] D
25. [4.4/352/1] A

FORM B

1. [4.1/327/3] A
2. [4.1/327/2] B
3. [4.1/327/3] B
4. [4.1/329/3] D
5. [4.2/331/3] A
6. [4.2/332/2] E
7. [4.2/335/2] C
8. [4.2/335/2] D
9. [4.2/333/1] A
10. [4.2/335/1] D
11. [4.2/336/2] B
12. [4.2/334/3] B
13. [4.3/341/1] A
14. [4.3/342/1] D
15. [4.3/342/1] D
16. [4.3/344/3] B
17. [4.3/344/2] A
18. [4.3/343/2] B
19. [4.3/344/2] C
20. [4.3/345/1] A
21. [4.3/345/1] C
22. [4.4/348/2] E
23. [4.4/350/1] C
24. [4.4/352/1] D
25. [4.4/352/1] E

FORM C

1. [4.1/327/3] D
2. [4.1/327/2] B
3. [4.1/327/3] B
4. [4.1/329/3] B
5. [4.2/331/3] E
6. [4.2/332/2] E
7. [4.2/335/2] C
8. [4.2/335/2] B
9. [4.2/333/1] B
10. [4.2/335/1] D

11.[4.2/336/2] C

12.[4.2/334/3] E

13.[4.3/341/1] C

14.[4.3/342/1] B

15.[4.3/342/1] A

16.[4.3/344/3] B

17.[4.3/344/2] A

18.[4.3/343/2] A

19.[4.3/344/2] C

20.[4.3/345/1] A

21.[4.3/345/1] C

22.[4.4/348/2] E

23.[4.4/350/1] C

24.[4.4/352/1] D

25.[4.4/352/1] E

FORM D

1. [4.1/327/3]

$2x^2+6+\frac{-5}{x+2}$

2. [4.1/327/2]

$x^3+2x^2+6x+17+\frac{44}{x-2}$

3. [4.1/327/3] 6

4. [4.1/329/3] $x-8$

5. [4.2/331/3] $6i$

6. [4.2/332/2] $-i$

7. [4.2/335/2] $5-2i$

8. [4.2/335/2] $35-30i$

9. [4.2/333/1]

$-1\pm i\sqrt{8}$

10.[4.2/335/1] $3-i$

11.[4.2/336/2] 12.0416

12.[4.2/334/3] $8+4i$

13.[4.3/341/1] any of

$x-5i$, $x+5i$, or $x+i$

14.[4.3/342/1]

$311+24i$

15.[4.3/342/1] $18+10i$

16.[4.3/344/3]

7 multiplicity 4;

–4 multiplicity 3

17.[4.3/344/2]

$-5i$ multiplicity 2;

$6-2i$ multiplicity 4

18.[4.3/343/2] –6, 3, 6

19.[4.3/344/2]

x^3-x^2-4x+4

20.[4.3/345/1]

x^3-x^2+4x-4

21.[4.3/345/1]

$a(x-7)(x-5-3i)(x-5+3i)$

22.[4.4/348/2]

–7, 1, 3, 7

23.[4.4/350/1]

$-9,-\frac{1}{2},\frac{1}{3},9$

24.[4.4/352/1]

$-\sqrt{7}$, 1, 3, $\sqrt{7}$

25.[4.4/352/1]

$-6i,-1,-3,6i$

FORM E

1. [4.1/327/3]

$2x^2+8+\frac{-9}{x+2}$

2. [4.1/327/2] B

$x^3+2x^2+6x+17+\frac{46}{x-2}$

3. [4.1/327/3] 10

4. [4.1/329/3] $x-9$

5. [4.2/331/3] $11i$

6. [4.2/332/2] $-i$

7. [4.2/335/2] $7-3i$

8. [4.2/335/2] $47+4i$

9. [4.2/333/1] $-1\pm\sqrt{5}$

10.[4.2/335/1] $\dfrac{11-3i}{2}$

11.[4.2/336/2] 7.6158

12.[4.2/334/3] $2+4i$

13.[4.3/341/1] any of

$x-6i$, $x+6i$, or $x+i$

14.[4.3/342/1]

$669+32i$

15.[4.3/342/1] $28+12i$

16.[4.3/344/3]

8 multiplicity 2;

–5 multiplicity 3

17.[4.3/344/2]

$-8i$ multiplicity 2;

$3-9i$ multiplicity 4

18.[4.3/343/2]

–12, 3, 12

19.[4.3/344/2]

x^3-x^2-4x+4

20.[4.3/345/1]

x^3-x^2+4x-4

21.[4.3/345/1]

$a(x-4)(x-5-2i)(x-5+2i)$

22.[4.4/348/2]

–10, 1, 3, 10

23.[4.4/350/1]

$-7, -\frac{1}{2}, \frac{1}{3}, 7$

24.[4.4/352/1]

$-\sqrt{6}$, 1, 3, $\sqrt{6}$

25.[4.4/352/1]

$-5i$, 1, 3, $5i$

FORM F

1. [4.1/327/3]

$2x^2+3+\dfrac{3}{x+2}$

2. [4.1/327/2]

$x^3+2x^2+6x+17+\dfrac{43}{x-2}$

3. [4.1/327/3] 15

4. [4.1/329/3] $x-2$

5. [4.2/331/3] $12i$

6. [4.2/332/2] $-i$

7. [4.2/335/2] $8-5i$

8. [4.2/335/2] $26-22i$

9. [4.2/333/1]

$-1\pm i\sqrt{6}$

10.[4.2/335/1] $8-5i$

11.[4.2/336/2] 8.0623

12.[4.2/334/3] $8+7i$

13.[4.3/341/1] any of

$x-7i$, $x+7i$, or $x+i$

14.[4.3/342/1]

$113+16i$

15.[4.3/342/1] $30+12i$

16.[4.3/344/3]

4 multiplicity 2;

–8 multiplicity 3

17.[4.3/344/2]

–7 multiplicity 2;

$8-2i$ multiplicity 4

18.[4.3/343/2] –8, 3, 8

19.[4.3/344/2]

x^3-x^2-4x+4

20.[4.3/345/1]

$x^3 - x^2 + 4x - 4$

21.[4.3/345/1]

$a(x-6)(x-7-3i)(x-7+3i)$

22.[4.4/348/2]

$-9, 1, 3, 9$

23.[4.4/350/1]

$-3, -\frac{1}{3}, \frac{1}{2}, 3$

24.[4.4/352/1]

$-\sqrt{11}, 1, 3, \sqrt{11}$

25.[4.4/352/1]

$-6i, 1, 3, 6i$

CHAPTER 5

FORM A

1. [5.1/361/2] A
2. [5.1/363/2] A
3. [5.1/363/2] C
4. [5.1/366/1] A
5. [5.2/375/1] B
6. [5.2/376/3] D
7. [5.2/376/2] D
8. [5.2/376/1] C
9. [5.2/377/2] C
10. [5.2/377/1] E
11. [5.2/378/1] C
12. [5.2/379/1] E
13. [5.3/385/2] D
14. [5.3/385/1] B
15. [5.3/385/3] C
16. [5.3/386/3] D
17. [5.3/386/3] B
18. [5.3/389/1] C
19. [5.3/389/1] E
20. [5.4/396/2] C
21. [5.4/396/2] D
22. [5.4/396/1] B
23. [5.4/397/1] C
24. [5.4/397/1] A
25. [5.5/405/1] E

FORM B

1. [5.1/361/2] B
2. [5.1/363/2] A
3. [5.1/363/2] A
4. [5.1/366/1] E
5. [5.2/375/1] D
6. [5.2/376/3] C
7. [5.2/376/2] C
8. [5.2/376/1] C
9. [5.2/377/2] B
10. [5.2/377/1] E
11. [5.2/378/1] D
12. [5.2/379/1] E
13. [5.3/385/2] B
14. [5.3/385/1] C
15. [5.3/385/3] B
16. [5.3/386/3] D
17. [5.3/386/3] C
18. [5.3/389/1] C
19. [5.3/389/1] E
20. [5.4/396/2] C
21. [5.4/396/2] A
22. [5.4/396/1] A
23. [5.4/397/1] C
24. [5.4/397/1] B
25. [5.5/405/1] E

FORM C

1. [5.1/361/2] C

2. [5.1/363/2] A

3. [5.1/363/2] B

4. [5.1/366/1] B

5. [5.2/375/1] A

6. [5.2/376/3] C

7. [5.2/376/2] C

8. [5.2/376/1] C

9. [5.2/377/2] A

10.[5.2/377/1] E

11.[5.2/378/1] A

12.[5.2/379/1] E

13.[5.3/385/2] B

14.[5.3/385/1] D

15.[5.3/385/3] A

16.[5.3/386/3] D

17.[5.3/386/3] A

18.[5.3/389/1] C

19.[5.3/389/1] E

20.[5.4/396/2] C

21.[5.4/396/2] B

22.[5.4/396/1] D

23.[5.4/397/1] C

24.[5.4/397/1] E

25.[5.5/405/1] E

FORM D

1. [5.1/361/2] 0.0156

2. [5.1/363/2]

Domain $(-\infty, \infty)$;

Range $(0, \infty)$

3. [5.1/363/2]

Horizontal shift 1 left;

Vertical shift 14 up

4. [5.1/366/1]

8 years

5. [5.2/375/1] 0.0008

6. [5.2/376/3] –8

7. [5.2/376/2] –10

8. [5.2/376/1] 9

9. [5.2/377/2] –1.2260

10.[5.2/377/1]

No solution

11.[5.2/378/1]

14650.74

12.[5.2/379/1]

14656.83

13.[5.3/385/2] $\log_4 x$

14.[5.3/385/1]

$x = \log_8 y$

15.[5.3/385/3] $x = 3^y$

16.[5.3/386/3] 4

17.[5.3/386/3] 0.8459

18.[5.3/389/1] $e^{x-3} + 7$

19.[5.3/389/1]

$\ln(x+3) - 7$

20.[5.4/396/2]

$7\log_3 x + 3\log_3 y - \log_3 z$

21.[5.4/396/2] $\log_4 2x$

22.[5.4/396/1] 8

23.[5.4/397/1]

8.7391

24.[5.4/397/1] 2.2756

25.[5.5/405/1] $3(8^x)$

FORM E

1. [5.1/361/2] 0.0078

2. [5.1/363/2]

Domain $(-\infty, \infty)$;

Range $(0, \infty)$

3. [5.1/363/2]

Horizontal shift 7 left;

Vertical shift 12 up

4. [5.1/366/1]

7 years

5. [5.2/375/1] 0.0006

6. [5.2/376/3] –7

7. [5.2/376/2] –2

8. [5.2/376/1] 9

9. [5.2/377/2] –0.4150

10.[5.2/377/1]

No solution

11.[5.2/378/1]

18313.43

12.[5.2/379/1]

18321.04

13.[5.3/385/2] $\log_8 x$

14.[5.3/385/1]

$x = \log_2 y$

15.[5.3/385/3] $x = 9^y$

16.[5.3/386/3] 5

17.[5.3/386/3] 0.9123

18.[5.3/389/1] $e^{x-4} + 9$

19.[5.3/389/1]

$\ln(x+4) - 9$

20.[5.4/396/2]

$5\log_5 x + 4\log_5 y - \log_5 z$

21.[5.4/396/2] $\log_3 x$

22.[5.4/396/1] 7

23.[5.4/397/1]

12.3182

24.[5.4/397/1] 0.9618

25.[5.5/405/1] $3(6^x)$

FORM F

1. [5.1/361/2] 0.0039

2. [5.1/363/2]

Domain $(-\infty, \infty)$;

Range $(0, \infty)$

3. [5.1/363/2]

Horizontal shift 9 left;

Vertical shift 19 up

4. [5.1/366/1]

9 years

5. [5.2/375/1] 0.0015

6. [5.2/376/3] –5

7. [5.2/376/2] –5

8. [5.2/376/1] 9

9. [5.2/377/2] –0.1072

10.[5.2/377/1]

No solution

11.[5.2/378/1] 9767.16

12.[5.2/379/1]

9771.22

13.[5.3/385/2] $\log_9 x$

14.[5.3/385/1]

$x = \log_6 y$

15.[5.3/385/3] $x = 11^y$

16.[5.3/386/3] 5

17.[5.3/386/3] 0.9858

18.[5.3/389/1] $e^{x-3} + 1$

19.[5.3/389/1]

$\ln(x+3) - 1$

20.[5.4/396/2]

$7\log_6 x + 6\log_6 y - \log_6 z$

21.[5.4/396/2] $\log_5 x$

22.[5.4/396/1] 3

23.[5.4/397/1]

13.4742

24.[5.4/397/1] 5.0230

25.[5.5/405/1] $3(2^x)$

CHAPTER 6

FORM A

1. [6.1/423/3] E

2. [6.1/423/1] E

3. [6.1/424/2] D

4. [6.1/424/2] B

5. [6.1/425/1] A

6. [6.1/425/3] E

7. [6.1/428/1] B

8. [6.2/437/3] A

9. [6.2/437/3] B

10.[6.2/437/2] B

11.[6.2/439/1] E

12.[6.2/440/1] E

13.[6.3/447/1] A

14.[6.3/448/1] E

15.[6.4/452/1] B

16.[6.4/452/1] B

17.[6.4/455/1] C

18.[6.5/460/2] B

19.[6.5/459/3] A

20.[6.5/461/2] C

21.[6.5/461/2] A

22.[6.5/461/1] A

23.[6.6/465/1] D

24.[6.6/465/1] D

25.[6.6/465/1] B

FORM B

1. [6.1/423/3] E

2. [6.1/423/1] E

3. [6.1/424/2] D

4. [6.1/424/2] C

5. [6.1/425/1] A

6. [6.1/425/3] E

7. [6.1/428/1] A

8. [6.2/437/3] A

9. [6.2/437/3] B

10.[6.2/437/2] A

11.[6.2/439/1] C

12.[6.2/440/1] C

13.[6.3/447/1] B

14.[6.3/448/1] E

15.[6.4/452/1] D

16.[6.4/452/1] B

17.[6.4/455/1] C

18.[6.5/460/2] C

19.[6.5/459/3] A

20.[6.5/461/2] C

21.[6.5/461/2] A

22.[6.5/461/1] A

23.[6.6/465/1] A

24.[6.6/465/1] B

25.[6.6/465/1] E

FORM C

1. [6.1/423/3] B

2. [6.1/423/1] B

3. [6.1/424/2] A

4. [6.1/424/2] E

5. [6.1/425/1] A

6. [6.1/425/3] E

7. [6.1/428/1] C

8. [6.2/437/3] A

9. [6.2/437/3] B

10.[6.2/437/2] C

11.[6.2/439/1] C

12.[6.2/440/1] D

13.[6.3/447/1] D

14.[6.3/448/1] E

15.[6.4/452/1] A

16.[6.4/452/1] B

17.[6.4/455/1] C

18.[6.5/460/2] B

19.[6.5/459/3] A

20.[6.5/461/2] C

21.[6.5/461/2] D

22.[6.5/461/1] A

23.[6.6/465/1] A

24.[6.6/465/1] B

25.[6.6/465/1] D

FORM D

1. [6.1/423/3] (1,2)

2. [6.1/423/1] (1,2)

3. [6.1/424/2] (2,1)

4. [6.1/424/2] (2, 1)

5. [6.1/425/1]

(2.6429, 1.0714)

6. [6.1/425/3]

No solution

7. [6.1/428/1]

4 quarters; 3 dimes

8. [6.2/437/3]

$$\begin{bmatrix} 7 & 2 & 5 \\ 4 & 3 & 7 \end{bmatrix}$$

9. [6.2/437/3]

$$\begin{aligned} 5x+4y&=6 \\ 2x+3y&=7 \end{aligned}$$

10.[6.2/437/2]

$$\begin{bmatrix} 1 & 0 & 4 & 3 \\ 0 & 1 & -8 & -3 \\ 1 & 7 & 2 & 6 \end{bmatrix}$$

11.[6.2/439/1]

(1, 3, 4)

12.[6.2/440/1]

$$y=\frac{1}{2}x^2+\frac{3}{2}x+8$$

13.[6.3/447/1]

(1, 1, −1)

14.[6.3/448/1]

No solution

15.[6.4/452/1]

(2, −5), (2,5)

16.[6.4/452/1]

(1, −1), (−1, 1)

17.[6.4/455/1]

$(-1, -\sqrt{8}), (1, -\sqrt{8}),$

$(-1, \sqrt{8}), (1, \sqrt{8})$

18.[6.5/460/2]

(1, 1), (1, 4), (0, 9)

19.[6.5/459/3]

$$-5x+y\le 3$$

20.[6.5/461/2]

$$\begin{aligned} -3x+y&\le 4 \\ -x+y&\le -5 \end{aligned}$$

21.[6.5/461/2] (1, 1)

22.[6.5/461/1]

$\left(0, \frac{4}{3}\right)$, (4, 0), (0, 0)

23.[6.6/465/1] 36

24.[6.6/465/1] (2, 2)

25.[6.6/465/1] $36

FORM E

1. [6.1/423/3] (1,2)

2. [6.1/423/1] (1,2)

3. [6.1/424/2] (2,1)

4. [6.1/424/2] (1,2)

5. [6.1/425/1]

(2.6429, 1.0714)

6. [6.1/425/3]

No solution

7. [6.1/428/1]

6 quarters; 3 dimes

8. [6.2/437/3]

$$\begin{bmatrix} 9 & 2 & 1 \\ 3 & 4 & 7 \end{bmatrix}$$

9. [6.2/437/3]

$$3x+9y=2$$
$$8x+7y=4$$

10.[6.2/437/2]

$$\begin{bmatrix} 1 & 0 & 1 & 1 \\ 0 & 1 & -3 & -2 \\ 6 & 8 & 1 & 7 \end{bmatrix}$$

11.[6.2/439/1]

(1, 3, 4)

12.[6.2/440/1]

$$y=\frac{1}{2}x^2+\frac{3}{2}x+10$$

13.[6.3/447/1]

(1, 1, −1)

14.[6.3/448/1]

No solution

15.[6.4/452/1]

No solution

16.[6.4/452/1]

(−1, 1), (1, −1)

17.[6.4/455/1]

$(-1, \sqrt{10}), (1, \sqrt{10}),$

$(-1, -\sqrt{10}), (1, -\sqrt{10})$

18.[6.5/460/2]

(1, 1), (1, 5), (0, 11)

19.[6.5/459/3]

$$-7x+y\le 6$$

20.[6.5/461/2]

$$-6x+y\le 4$$
$$-x+y\le -7$$

21.[6.5/461/2] (1,1)

22.[6.5/461/1]

$\left(0, \frac{4}{3}\right)$, (4, 0), (0, 0),

23.[6.6/465/1] 44

24.[6.6/465/1] (2, 2)

25.[6.6/465/1] $44

FORM F

1. [6.1/423/3] (1, 2)

2. [6.1/423/1] (1, 2)

3. [6.1/424/2] (2, 1)

4. [6.1/424/2] (1, 2)

5. [6.1/425/1]

(2.6429, 1.0714)

6. [6.1/425/3]

No solution

7. [6.1/428/1]

6 quarters; 4 dimes

8. [6.2/437/3]

$$\begin{bmatrix} 2 & 7 & 3 \\ 4 & 5 & 1 \end{bmatrix}$$

9. [6.2/437/3]

$$9x+7y=8$$
$$3x+6y=5$$

10.[6.2/437/2]

$$\begin{bmatrix} 1 & 0 & 2 & 1 \\ 0 & 1 & -10 & -4 \\ 3 & 7 & 2 & 6 \end{bmatrix}$$

11.[6.2/439/1]

(1, 3, 4)

12.[6.2/440/1]

$y=\frac{1}{2}x^2+\frac{3}{2}x+2$

13.[6.3/447/1]

(1, 1, −1)

14.[6.3/448/1]

No solution

15.[6.4/452/1]

(1, −5), (1, 5)

16.[6.4/452/1]

(1, −1), (−1, 1)

17.[6.4/455/1]

(−1, −1), (−1, 1),

(1, −1), (1, 1)

18.[6.5/460/2]

(1, 1), (1, 7), (0, 15)

19.[6.5/459/3]

$-7x+y\le 6$

20.[6.5/461/2]

$-6x+y\ge 4$
$-x+y\ge 7$

21.[6.5/461/2]

(1, 1)

22.[6.5/461/1]

(4, 0), $\left(0, \frac{4}{3}\right)$, (0, 0)

23.[6.6/465/1] 32

24.[6.6/465/1] (2, 2)

25.[6.6/465/1] $32

CHAPTER 7

FORM A

1. [7.1/478/3] E
2. [7.1/479/3] D
3. [7.1/480/2] A
4. [7.1/483/2] D
5. [7.1/479/3] A
6. [7.2/493/2] A
7. [7.2/496/2] C
8. [7.2/497/3] E
9. [7.3/503/3] C
10. [7.3/505/2] B
11. [7.3/505/1] D
12. [7.3/5071/1] C
13. [7.3/507/1] A
14. [7.3/509/1] B
15. [7.4/513/3] A
16. [7.4/515/2] D
17. [7.4/514/3] D
18. [7.4/517/2] C
19. [7.4/517/1] E
20. [7.4/518/1] E
21. [7.4/520/3] B
22. [7.4/520/3] A
23. [7.5/522/3] A
24. [7.5/523/3] B
25. [7.5/526/2] B

FORM B

1. [7.1/478/3] E
2. [7.1/479/3] E
3. [7.1/480/2] A
4. [7.1/483/2] D
5. [7.1/479/3] B
6. [7.2/493/2] B
7. [7.2/496/2] D
8. [7.2/497/3] E
9. [7.3/503/3] C
10. [7.3/505/2] A

11.[7.3/505/1] A

12.[7.3/507/1] A

13.[7.3/507/1] A

14.[7.3/509/1] C

15.[7.4/513/3] B

16.[7.4/515/2] D

17.[7.4/514/3] E

18.[7.4/517/2] C

19.[7.4/517/1] E

20.[7.4/518/1] E

21.[7.4/520/3] A

22.[7.4/520/3] A

23.[7.5/522/3] B

24.[7.5/523/3] B

25.[7.5/526/2] B

FORM C

1. [7.1/478/3] E

2. [7.1/479/3] A

3. [7.1/480/2] A

4. [7.1/483/2] A

5. [7.1/479/3] C

6. [7.2/493/2] D

7. [7.2/496/2] A

8. [7.2/497/3] E

9. [7.3/503/3] C

10.[7.3/505/2] C

11.[7.3/505/1] B

12.[7.3/507/1] D

13.[7.3/507/1] A

14.[7.3/509/1] E

15.[7.4/513/3] C

16.[7.4/515/2] D

17.[7.4/514/3] C

18.[7.4/517/2] C

19.[7.4/517/1] E

20.[7.4/518/1] E

21.[7.4/520/3] D

22.[7.4/520/3] A

23.[7.5/522/3] E

24.[7.5/523/3] B

25.[7.5/526/2] A

FORM D

1. [7.1/478/3] 2×3

2. [7.1/479/3] 3

3. [7.1/480/2]

(1, 3, 8)

4. [7.1/483/2]

$$\begin{bmatrix} 15 & 1 \\ 4 & 7 \end{bmatrix}$$

5. [7.1/479/3] 3

6. [7.2/493/2]

$$\begin{bmatrix} 2 & 22 \\ 1 & 10 \end{bmatrix}$$

7. [7.2/496/2] 2×2

8. [7.2/497/3]

Does not exist.

9. [7.3/503/3]

$$\begin{bmatrix} 1 & 0 & 0 \\ 0 & 1 & 0 \\ 0 & 0 & 1 \end{bmatrix}$$

10.[7.3/505/2]

$$\begin{bmatrix} 1 & -7 \\ 0 & 1 \end{bmatrix}$$

11.[7.3/505/1]

$$\begin{bmatrix} -.037 & .222 & -.481 \\ .074 & -.444 & 1.963 \\ .111 & .333 & -1.556 \end{bmatrix}$$

12.[7.3/507/1] (3, 2)

13.[7.3/507/1]

(3, 2, 1)

14.[7.3/509/1]

$$\begin{bmatrix} 930 \\ 858 \end{bmatrix}$$

15.[7.4/513/3] 11

16.[7.4/515/2] −132

17.[7.4/514/3] 0

18.[7.4/517/2] (3, 4)

19.[7.4/517/1]

Cramer's rule cannot be used.

20.[7.4/518/1]

Cramer's rule cannot be used.

21.[7.4/520/3] 6

22.[7.4/520/3] 0

23.[7.5/522/3]

$$\begin{bmatrix} 6 & 5 \\ 2 & 7 \end{bmatrix}$$

24.[7.5/523/3] 7

25.[7.5/526/2] 0

FORM E

1. [7.1/478/3] 2×3

2. [7.1/479/3] 4

3. [7.1/480/2]

(3, 1, 7)

4. [7.1/483/2]

$$\begin{bmatrix} 9 & 4 \\ 4 & 5 \end{bmatrix}$$

5. [7.1/479/3] 3

6. [7.2/493/2]

$$\begin{bmatrix} 15 & 11 \\ 5 & 7 \end{bmatrix}$$

7. [7.2/496/2] 5×5

8. [7.2/497/3]

Does not exist.

9. [7.3/503/3]

$$\begin{bmatrix} 1 & 0 & 0 \\ 0 & 1 & 0 \\ 0 & 0 & 1 \end{bmatrix}$$

10.[7.3/505/2]

$$\begin{bmatrix} 1 & -9 \\ 0 & 1 \end{bmatrix}$$

11.[7.3/505/1]

$$\begin{bmatrix} .029 & -.714 & .314 \\ .171 & .214 & -.114 \\ -.086 & .143 & .057 \end{bmatrix}$$

12.[7.3/507/1] (1, 5)

13.[7.3/507/1]

(3, 2, 1)

14.[7.3/509/1]

$$\begin{bmatrix} 677 \\ 648 \end{bmatrix}$$

15.[7.4/513/3] 26

16.[7.4/515/2] −2

17.[7.4/514/3] 0

18.[7.4/517/2] (3, 4)

19.[7.4/517/1]

Cramer's rule cannot be used.

20.[7.4/518/1]

Cramer's rule cannot be used.

21.[7.4/520/3] 34

22.[7.4/520/3] 0

23.[7.5/522/3]

$$\begin{bmatrix} 9 & 7 \\ 8 & 4 \end{bmatrix}$$

24.[7.5/523/3] 8

25.[7.5/526/2] 1

FORM F

1. [7.1/478/3] 2×3

2. [7.1/479/3] 1

3. [7.1/480/2]

(7, 2, 1)

4. [7.1/483/2]

$$\begin{bmatrix} 9 & 3 \\ 9 & 6 \end{bmatrix}$$

5. [7.1/479/3] 3

6. [7.2/493/2]

$$\begin{bmatrix} 24 & 44 \\ 21 & 27 \end{bmatrix}$$

7. [7.2/496/2] 5×5

8. [7.2/497/3]

Does not exist.

9. [7.3/503/3]

$$\begin{bmatrix} 1 & 0 & 0 \\ 0 & 1 & 0 \\ 0 & 0 & 1 \end{bmatrix}$$

10.[7.3/505/2]

$$\begin{bmatrix} 1 & -12 \\ 0 & 1 \end{bmatrix}$$

11.[7.3/505/1]

$$\begin{bmatrix} .214 & -.357 & -.071 \\ .786 & -3.643 & .071 \\ -.429 & 1.714 & .143 \end{bmatrix}$$

12.[7.3/507/1] (1, 3)

13.[7.3/507/1]

(3, 2, 1)

14.[7.3/509/1]

$$\begin{bmatrix} 482 \\ 548 \end{bmatrix}$$

15.[7.4/513/3] 44

16.[7.4/515/2] 48

17.[7.4/514/3] 0

18.[7.4/517/2] (3, 4)

19.[7.4/517/1]

Cramer's rule cannot be used.

20.[7.4/518/1]

Cramer's rule cannot be used.

21.[7.4/520/3] 27

22.[7.4/520/3] 0

23.[7.5/522/3]

$$\begin{bmatrix} 4 & 3 \\ 7 & 6 \end{bmatrix}$$

24.[7.5/523/3] 7

25.[7.5/526/2] 1

CHAPTER 8

FORM A

1. [8.1/537/3] D

2. [8.1/537/3] B

3. [8.1/538/3] A

4. [8.1/540/2] B

5. [8.1/539/2] E

6. [8.1/540/2] A

7. [8.1/540/2] D

8. [8.2/545/3] B

9. [8.2/545/3] D

10.[8.2/545/3] D

11.[8.2/545/2] C

12.[8.2/546/2] C

13.[8.2/547/2] C

14.[8.2/548/1] D

15.[8.3/553/3] B

16.[8.2/553/2] C

17.[8.3/556/1] A

18.[8.3/558/2] A

19.[8.3/558/3] E

20.[8.4/567/1] D

21.[8.5/572/1] B

22.[8.5/574/3] A

23.[8.5/574/3] E

24.[8.5/575/1] B

25.[8.5/575/1] B

FORM B

1. [8.1/537/3] B

2. [8.1/537/3] A

3. [8.1/538/3] A

4. [8.1/540/2] A

5. [8.1/539/2] A

6. [8.1/540/2] C

7. [8.1/540/2] A

8. [8.2/545/3] A

9. [8.2/545/3] A

10.[8.2/545/3] C

11.[8.2/545/2] B

12.[8.2/546/2] B

13.[8.2/547/2] B

14.[8.2/548/1] D

15.[8.3/553/3] B

16.[8.2/553/2] C

17.[8.3/556/1] A

18.[8.3/558/2] B

19.[8.3/558/3] B

20.[8.4/567/1] C

21.[8.5/572/1] B

22.[8.5/574/3] D

23.[8.5/574/3] C

24.[8.5/575/1] C

25.[8.5/575/1] C

FORM C

1. [8.1/537/3] E

2. [8.1/537/3] E

3. [8.1/538/3] A

4. [8.1/540/2] D

5. [8.1/539/2] B

6. [8.1/540/2] B

7. [8.1/540/2] B

8. [8.2/545/3] E

9. [8.2/545/3] B

10.[8.2/545/3] C

11.[8.2/545/2] D

12.[8.2/546/2] D

13.[8.2/547/2] D

14.[8.2/548/1] E

15.[8.3/553/3] B

16.[8.2/553/2] C

17.[8.3/556/1] B

18.[8.3/558/2] D

19.[8.3/558/3] A

20.[8.4/567/1] D

21.[8.5/572/1] B

22.[8.5/574/3] A

23.[8.5/574/3] A

24.[8.5/575/1] D

25.[8.5/575/1] D

FORM D

1. [8.1/537/3]

10, 14, 18, 22

2. [8.1/537/3] 112

3. [8.1/538/3]

5, 17, 29, 41, 53

4. [8.1/540/2] 95

5. [8.1/539/2]

$$\sum_{k=1}^{4}(8+k)$$

6. [8.1/540/2] 33

7. [8.1/540/2] 17

8. [8.2/545/3] 5

9. [8.2/545/3] 24, 30

10.[8.2/545/3]

9, 12, 15, 18

11.[8.2/545/2] 30

12.[8.2/546/2] 30

13.[8.2/547/2] 30

14.[8.2/548/1] 195

15.[8.3/553/3] –2

16.[8.2/553/2]

$4(3)^{14}$

17.[8.3/556/1]

5.5556

18.[8.3/558/2] 12

19.[8.3/558/3] $\frac{734}{990}$

20.[8.4/567/1]

$5n(n+1)$

21.[8.5/572/1]

$a^5 + 20a^4b + 160a^3b^2 + 640a^2b^3 + 1280ab^4 + 1024b^5$

22.[8.5/574/3]

8,742

23.[8.5/574/3] 36

24.[8.5/575/1]

$70x^4y^4$

25.[8.5/575/1]

$70x^4y^4$

FORM E

1. [8.1/537/3]

7, 11, 15, 19

2. [8.1/537/3] 207

3. [8.1/538/3]

5, 11, 17, 23, 29

4. [8.1/540/2] 95

5. [8.1/539/2]

$\sum_{k=1}^{4}(11+k)$

6. [8.1/540/2] 21

7. [8.1/540/2] 25

8. [8.2/545/3] 6

9. [8.2/545/3] 36, 45

10.[8.2/545/3]

5, 13, 21, 29

11.[8.2/545/2] 36

12.[8.2/546/2] 36

13.[8.2/547/2] 36

14.[8.2/548/1] 162

15.[8.3/553/3] –2

16.[8.2/553/2] $4(3)^{15}$

17.[8.3/556/1]

7.8750

18.[8.3/558/2]

11.6667

19.[8.3/558/3] $\frac{833}{990}$

20.[8.4/567/1]

$6n(n+1)$

21.[8.5/572/1]

$a^5 - 20a^4b + 160a^3b^2 - 640a^2b^3 + 1280ab^4 - 1024b^5$

22.[8.5/574/3]

970,200

23.[8.5/574/3] 36

24.[8.5/575/1]

$28x^6y^2$

25.[8.5/575/1]

$28x^6y^2$

FORM F

1. [8.1/537/3]

9, 13, 17, 21

2. [8.1/537/3] 309

3. [8.1/538/3]

5, 14, 23, 32, 41

4. [8.1/540/2] 115

5. [8.1/539/2]

$\sum_{k=1}^{4}(16+k)$

6. [8.1/540/2] 33

7. [8.1/540/2] 3

8. [8.2/545/3] 5

9. [8.2/545/3] 40, 50

10.[8.2/545/3]

7, 11, 15, 19

11.[8.2/545/2] 42

12.[8.2/546/2] 42

13.[8.2/547/2] 42

14.[8.2/548/1] 162

15.[8.3/553/3] –2

16.[8.2/553/2] $4(3)^{19}$

17.[8.3/556/1]

10.9091

18.[8.3/558/2]

10.2857

19.[8.3/558/3] $\dfrac{932}{990}$

20.[8.4/567/1]

$7n(n+1)$

21.[8.5/572/1]

$1024a^5+1280a^4b+640a^3b^2+$
$160a^2b^3+20ab^4+b^5$

22.[8.5/574/3]

10,302

23.[8.5/574/3] 210

24.[8.5/575/1]

$9x^8y$

25.[8.5/575/1]

$-9x^8y$

CHAPTER 9

FORM A

1. [9.1/587/3] D

2. [9.1/587/3] C

3. [9.1/589/3] A

4. [9.1/585/3] E

5. [9.1/585/3] B

6. [9.1/587/2] A

7. [9.1/587/2] B

8. [9.1/589/1] D

9. [9.1/589/1] D

10.[9.2/594/3] D

11.[9.2/601/1] C

12.[9.2/594/1] C

13.[9.2/595/1] C

14.[9.2/597/1] E

15.[9.2/597/1] D

16.[9.2/597/1] B

17.[9.2/597/1] B

18.[9.2/599/2] C

19.[9.2/602/1] B

20.[9.2/602/1] A

21.[9.2/603/1] B

22.[9.2/603/1] A

23.[9.1/587/3] D

24.[9.1/587/3] E

25.[9.1/589/3] E

FORM B

1. [9.1/587/3] A

2. [9.1/587/3] A

3. [9.1/589/3] D

4. [9.1/585/3] A

5. [9.1/585/3] C

6. [9.1/587/2] A

7. [9.1/587/2] B

8. [9.1/589/1] D

9. [9.1/589/1] D

10.[9.2/594/3] D

11.[9.2/601/1] C

12.[9.2/594/1] B

13.[9.2/595/1] D

14.[9.2/597/1] E

15.[9.2/597/1] D

16.[9.2/597/1] A

17.[9.2/597/1] C

18.[9.2/599/2] C

19.[9.2/602/1] B

20.[9.2/602/1] C

21.[9.2/603/1] A

22.[9.2/603/1] A

23.[9.1/587/3] C

24.[9.1/587/3] E

25.[9.1/589/3] E

FORM C

1. [9.1/587/3] C

2. [9.1/587/3] D

3. [9.1/589/3] B

4. [9.1/585/3] B

5. [9.1/585/3] C

6. [9.1/587/2] A

7. [9.1/587/2] B

8. [9.1/589/1] D

9. [9.1/589/1] D

10.[9.2/594/3] D

11.[9.2/601/1] C

12.[9.2/594/1] C

13.[9.2/595/1] B

14.[9.2/597/1] E

15.[9.2/597/1] D

16.[9.2/597/1] C

17.[9.2/597/1] D

18.[9.2/599/2] C

19.[9.2/602/1] B

20.[9.2/602/1] E

21.[9.2/603/1] D

22.[9.2/603/1] A

23.[9.1/587/3] E

24.[9.1/587/3] C

25.[9.1/589/3] C

FORM D

1. [9.1/587/3]

479,001,600

2. [9.1/587/3] 210

3. [9.1/589/3] 15

4. [9.1/585/3] 48

5. [9.1/585/3] 36

6. [9.1/587/2]

8! or 40,320

7. [9.1/587/2]

${}_{10}P_3$ or 720

8. [9.1/589/1]

${}_8C_5$ or 56

9. [9.1/589/1]

${}_8C_5$ or 56

10.[9.2/594/3] $\frac{1}{4}$

11.[9.2/601/1]

$\frac{{}_{12}C_4}{2^{12}}$ or 0.1208

12.[9.2/594/1] $\frac{1}{13}$

13.[9.2/595/1] $\frac{4}{36}$

14.[9.2/597/1] 0.0121

15.[9.2/597/1] $\frac{4}{13}$

16.[9.2/597/1] $\frac{1}{52}$

17.[9.2/597/1] $\frac{2}{13}$

18.[9.2/599/2] $\frac{12}{25}$

19.[9.2/602/1]

$\frac{{}_{48}C_1}{{}_{52}C_5}$ or 0.00001847

20.[9.2/602/1]

$\frac{{}_{40}C_2\ {}_5C_1}{{}_{45}C_3}$ or 0.2748

21.[9.2/603/1]

$\frac{{}_{40}C_2\ {}_5C_1}{{}_{45}C_3}+\frac{{}_{40}C_3}{{}_{45}C_3}$

22.[9.2/603/1]

$1-\frac{{}_{40}C_2\ {}_5C_1}{{}_{45}C_3}+\frac{{}_{40}C_3}{{}_{45}C_3}$

23.[9.1/587/3] 6

24.[9.1/587/3] 1

25.[9.1/589/3] 1

FORM E

1. [9.1/587/3] 120

2. [9.1/587/3] 360

3. [9.1/589/3] 12

4. [9.1/585/3] 80

5. [9.1/585/3] 28

6. [9.1/587/2]

9! or 362,880

7. [9.1/587/2]

${}_9P_3$ or 504

8. [9.1/589/1]

${}_9C_6$ or 84

9. [9.1/589/1]

${}_9C_6$ or 84

10.[9.2/594/3] $\frac{1}{4}$

11.[9.2/601/1]

$\frac{{}_{16}C_5}{2^{16}}$ or 0.0667

12.[9.2/594/1] $\frac{1}{13}$

13.[9.2/595/1] $\frac{2}{36}$

14.[9.2/597/1] 0.0121

15.[9.2/597/1] $\frac{7}{13}$

16.[9.2/597/1] $\frac{1}{52}$

17.[9.2/597/1] $\frac{2}{13}$

18.[9.2/599/2] $\frac{1}{2}$

19.[9.2/602/1]

$\frac{{}_{48}C_1}{{}_{52}C_5}$ or 0.00001847

20.[9.2/602/1]

$\frac{{}_{30}C_2\ {}_6C_1}{{}_{36}C_3}$ or 0.3655

21.[9.2/603/1]

$\frac{{}_{30}C_2\ {}_6C_1}{{}_{36}C_3}+\frac{{}_{30}C_3}{{}_{36}C_3}$

22.[9.2/603/1]

$1-\frac{{}_{30}C_2\ {}_6C_1}{{}_{36}C_3}-\frac{{}_{30}C_3}{{}_{36}C_3}$

23.[9.1/587/3] 10

24.[9.1/587/3] 1

25.[9.1/589/3] 1

FORM F

1. [9.1/587/3] 40,320

2. [9.1/587/3] 504

3. [9.1/589/3] 7

4. [9.1/585/3] 48

5. [9.1/585/3] 45

6. [9.1/587/2]

11! or 39,916,800

7. [9.1/587/2]

${}_8P_3$ or 336

8. [9.1/589/1]

${}_8C_3$ or 56

9. [9.1/589/1]

${}_8C_3$ or 56

10.[9.2/594/3] $\frac{1}{4}$

11.[9.2/601/1]

$\frac{{}_{11}C_4}{2^{11}}$ or 0.1611

12.[9.2/594/1] $\frac{1}{4}$

13.[9.2/595/1] $\frac{2}{36}$

14.[9.2/597/1] 0.0121

15.[9.2/597/1] $\frac{7}{13}$

16.[9.2/597/1] $\frac{1}{52}$

17.[9.2/597/1] $\frac{2}{13}$

18.[9.2/599/2] $\frac{8}{19}$

19.[9.2/602/1]

$\frac{{}_{48}C_1}{{}_{52}C_5}$ or 0.00001847

20.[9.2/602/1]

$\frac{{}_{45}C_2\ {}_5C_1}{{}_{50}C_3}$ or 0.2526

21.[9.2/603/1]

$\frac{{}_{45}C_2\ {}_5C_1}{{}_{50}C_3}+\frac{{}_{45}C_3}{{}_{50}C_3}$

22.[9.2/603/1]

$1-\frac{{}_{45}C_2\ {}_5C_1}{{}_{50}C_3}-\frac{{}_{45}C_3}{{}_{50}C_3}$

23.[9.1/587/3] 13

24.[9.1/587/3] 1

25.[9.1/589/3] 1

FINAL EXAM

FORM A

1. A
2. C
3. E
4. A
5. D
6. B
7. B
8. A
9. C
10. C
11. B
12. D
13. C
14. C
15. C
16. B
17. A
18. D
19. C
20. D
21. B
22. D
23. D

24. B

25. A

26. A

27. A

28. E

29. A

30. B

31. B

32. A

33. A

34. C

35. A

36. C

37. E

38. B

39. D

40. A

FORM B

1. A

2. D

3. D

4. A

5. E

6. A

7. C

8. D

9. D

10. C

11. A

12. D

13. C

14. E

15. C

16. D

17. C

18. B

19. C

20. A

21. E

22. E

23. C

24. C

25. D

26. B

27. A

28. C

29. A

30. A

31. B

32. A

33. C

34. C

35. B

36. C

37. A

38. B

39. D

40. A

FORM C

1. x-intercept (10, 0); y-intercept (0, 5)

2. $y-10=\frac{25}{7}(x-7)$

3. (−1.5, −16.5)

4. (−1.1832, −4.2), (1.1832, −4.2)

5. $(x-8)^2+(y+12)^2=36$

6. −16, −4

7. 18 seconds

8. 52

9. $(7, \infty)$

10. $(-32, 32)$

11. Domain: $(-\infty,\infty)$
Range: $[7,\infty]$

12. All real numbers except 4

13. $x^8 - 2x^4$

14. $g(x)$ only

15. $\sqrt[7]{\frac{x+4}{3}}$

16. $36+33i$

17. $1-i$

18. $9-6i$

19. $-2, 2, 3, 5$

20. -2

21. -1.7897

22. 7

23. 15

24. 16.1181

25. 10 quarters;
20 nickels

26. $\begin{bmatrix} 1 & 2 & 2 & 1 \\ 4 & 0 & 2 & 6 \\ 1 & 1 & 3 & 1 \end{bmatrix}$

27. $(1, 2, 2)$

28. $84,000

29. $(5, 7, 2)$

30. $\begin{bmatrix} 144 & 24 \\ 144 & 25 \end{bmatrix}$

31. -2

32. 80

33. $8, 5, 2, -1$

34. 36

35. $9(5)^7$

36.
$a^5 - 10a^4b + 40a^3b^2 - 80a^2b^3 + 80ab^4 - 32b^5$

37. 427,518,000

38. ${}_{90}P_3$

39. $\frac{1}{52}$

40. $\frac{{}_{96}C_2 \; {}_4C_1 + {}_{96}C_3}{{}_{100}C_3}$

FORM D

1. x-intercept $(6, 0)$;
y-intercept $(0, 4)$

2. $y - 10 = 5(x - 10)$

3. $(-1.5, -24)$

4. $(-1.0690, -3.4286)$,
$(1.0690, -3.4286)$

5.
$(x-9)^2 + (y+11) = 49$

6. 4, 16

7. 20 seconds

8. 88

9. $(9, \infty)$

10. $(-51, 51)$

11. Domain: $(-\infty, \infty)$;
Range: $[4,\infty)$

12. All real numbers except 5

13. $x^{15} + 3x^5$

14. $h(x)$ only

15. $\sqrt[11]{\frac{x+6}{5}}$

16. $-2+26i$

17. $1+i$

18. $11-4i$

19. $-2, 2, 3, 6$

20. -2

21. −0.4599

22. 6

23. 7

24. 13.1833

25. 10 quarters;
22 nickels

26. $\begin{bmatrix} 1 & 2 & 2 & 1 \\ 5 & 1 & -1 & 3 \\ 1 & 4 & 1 & 3 \end{bmatrix}$

27. (2, 2, 1)

28. $252,000

29. (3, 2, 4)

30. $\begin{bmatrix} 66 & 82 \\ 57 & 72 \end{bmatrix}$

31. −9

32. 100

33. 7, 2, −3, −8

34. 32

35. $8(9)^7$

36.
$32a^5 - 80a^4b + 80a^3b^2$
$- 40a^2b^3 + 10ab^4 - b^5$

37. 657,720

38. ${}_{78}P_3$

39. $\frac{1}{52}$

40. $\frac{{}_{95}C_2 \; {}_{5}C_1 + {}_{95}C_3}{{}_{100}C_3}$